The Cambridge Manuals of Science and
Literature

THE   MODERN   LOCOMOTIVE

'The Old and the New.' Latest express 4–4–0 type superheater passenger engine of the London and North Western Railway alongside the *Rocket*.

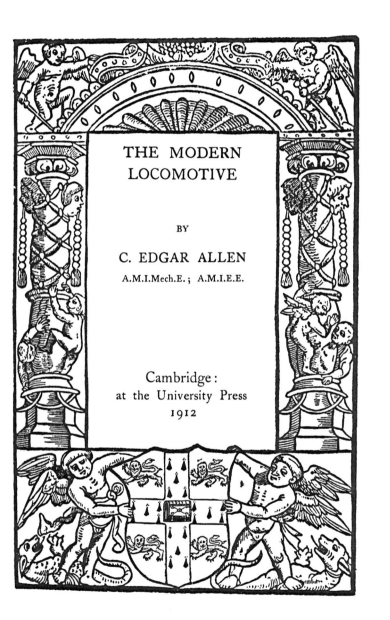

THE MODERN LOCOMOTIVE

BY

C. EDGAR ALLEN
A.M.I.Mech.E.; A.M.I.E.E.

Cambridge:
at the University Press
1912

CAMBRIDGE UNIVERSITY PRESS
Cambridge, New York, Melbourne, Madrid, Cape Town,
Singapore, São Paulo, Delhi, Tokyo, Mexico City

Cambridge University Press
The Edinburgh Building, Cambridge CB2 8RU, UK

Published in the United States of America by
Cambridge University Press, New York

www.cambridge.org
Information on this title: www.cambridge.org/9781107401709

© Cambridge University Press 1912

First published 1912
First paperback edition 2011

A catalogue record for this publication is available from the British Library

ISBN 978-1-107-40170-9 Paperback

*With the exception of the coat of arms
at the foot, the design on the title page is a
reproduction of one used by the earliest known
Cambridge printer, John Siberch, 1521*

PREFACE

IN a small book, not intended for specialists but for a wider public, it has not been possible to do more than sketch the general principles governing the design and working of a modern locomotive, and to trace the broad lines of development from its comparatively simple predecessor of twenty-five or thirty years ago. More attention has been given to such matters as combustion, transfer of heat, steam production, superheating, compounding, feed-water heating, resistance and stability, as being essential to the proper understanding of the modern locomotive, than to mechanical features, to do justice to which would involve a mass of technical detail. That phase of the subject which catalogues the dimensions of various types of engines actually in service has also been avoided.

The author appreciates his indebtedness to many of the books and proceedings mentioned in the bibliography, and he desires to acknowledge the information supplied by Mr J. G. Bowen Cooke, Chief Mechanical Engineer of the L. & N. W. Railway, and Mr W. P. Reid, Chief Mechanical Engineer of the North British Railway. His thanks are due to the Editor of the *Engineer* for permission to reproduce the illustration on page 24. In a similar manner he is indebted to Messrs Constable & Co., Messrs Doin et Fils, Paris, and to that valuable work, *The Locomotive of To-day*, for the illustrations on pages 55, 109 and 43 respectively.

C. E. A.

WATFORD,
  December, 1911.

CONTENTS

THE MODERN LOCOMOTIVE

INTRODUCTION

FEW subjects possess more importance for the public to-day than travelling, and it is safe to say that in the mind of the majority this is primarily associated with the railway and more vaguely with the locomotive. Long familiarity with the locomotive, coupled with a general sameness in its external appearance, has engendered indifference and rendered all who have not given special attention to the subject oblivious to the fact that many and radical changes have been taking place in its design. Moreover its hold on public interest has to some extent been challenged by a younger rival, the electric locomotive or electrically propelled train, whose possibilities are generally credited with an exaggerated importance. Many no longer express astonishment at the tale of the achievement of the 'iron horse,' but are inclined to call into question its easy pre-eminence for hauling fast and heavy trains. If put to it however, they will confess to a sentimental interest aroused by the attributes of strength, symmetry and self-sufficiency possessed by the locomotive, and would confess to a pang of regret if it were ultimately displaced in the fierce conflict of less noble, if not less efficient, competitive systems.

In spite of all that is heard to the contrary, one feature stands out very clearly in the mind of those qualified to form an accurate judgment, that is, there is no reasonable prospect of our trusted friend being relegated to the scrap heap or becoming an isolated relic allowed to stand in solitary grandeur on a concrete foundation in leading railway stations. The limitations of electric traction are far too clearly recognized and easily defined, and unless some un-foreseen and revolutionary change takes place the steam locomotive will be found doing its duty many years hence. Here it may be stated that the true object of the electrification of railways is the diversion of passengers from competing tramways or omnibuses, or the development of populous districts, both of which are local problems and have little in common with the problem of the steam working of main-line traffic.

For the reason of the existence of the locomotive we must look back far into the seventeenth century, when, in the remote colliery districts of the North, coal haulage was laboriously effected by wagons slowly dragged along wooden tramroads by horses. The feeble resistance of wooden rails to wear and their susceptibility to decay led to their displacement by rails made of iron, which were widely adopted in most of the colliery districts.

While horse-power or the stationary steam engine remained the only tractive force available for the haulage of wagons, a fixed limit was placed on the development of the railway system. It was the invention of the locomotive by Trevithick and its subsequent improvement by the Stephensons, that was to give a great impetus to the construction of iron roads, and place the question of steam locomotion on a successful basis. It is only when we hold steadily in view the position occupied by railways to-day as compared with their humble origin of a century ago, that we are enabled to realise how much we owe to the locomotive and how greatly economic history is bound up with it.

The writer's intention is not to cover the ground of more or less ancient history and examine in detail the gradual development of the locomotive to its present dimensions and design; it will suffice to say that the original engines of the Stockton and Darlington or the Liverpool and Manchester Railways would be regarded as mere toys compared with a modern express engine. Their relative size and capacity are well illustrated by the photograph

1—2

reproduced in the frontispiece to this volume shewing
the *Rocket*, or rather an exact reproduction of that
famous engine, alongside one of Mr Bowen Cooke's
latest express engines on the London and North
Western Railway.

It was in the middle fifties that design settled
down to the definite type with the four coupled
driving wheels and a leading pair of carrying
wheels, which was to remain for so long the standard
practice of Great Britain and the Continent. This
wheel arrangement, with the exception that a four-
wheel bogie has replaced the pair of leading wheels,
may be said to predominate even to-day. From the
period mentioned until 1870 nothing very remarkable
in the way of new development took place, but the
ensuing twenty-five years witnessed the introduction
of many fresh and notable locomotive types. It is
a tribute to the excellence of those engines that a
number of them—witness the *Precedent* class of the
London and North Western Railway, rebuilt it is true
—are still running. The *Charles Dickens* (No. 955),
the most famous of this class, left the Crewe shops
in 1882 and ran daily from Manchester to Euston
and back, a distance of 366½ miles, completing her
millionth mile 9 years 219 days after. During
that period she made 2651 runs between the towns
mentioned in addition to 92 other journeys, and
consumed 12,515 tons of coal.

Charles Dickens continued to work the 8.30 a.m.
Manchester to London and 4.0 p.m. London to Manchester trains until August 5th, 1902, on which date
the engine completed 2,000,000 miles on the return
journey from London to Manchester, having run
5312 trips to London and back, in addition to 186
other sundry trips. The engine was withdrawn from
the Manchester to Euston express passenger service
on August 5th, 1902 : it has since been working
passenger trains between Manchester, Crewe, Shrews-
bury, Birmingham and Leeds, and has occasionally
run goods trains. The total number of miles run
by this engine from the date of first turning out of
the works in February, 1882, up to and including
October 31st, 1911, was 2,318,918.

Other notable engines of this period were
Mr D. Drummond's four-coupled engines on the
Caledonian ; Mr W. Stroudley's *Gladstones* on the
London, Brighton and South Coast ; Mr S. W. Johnson's
single wheelers on the Midland, and Mr Patrick
Stirling's 8 ft. singles on the Great Northern.

But if the period above mentioned witnessed the
production of a number and variety of new types
what must be said of the last twelve years ? The
speed of trains has materially increased, the demand
for the most rapid transport of both passenger and
goods has become more urgent, and, further, the
taste of the public coupled with competition amongst

the railway companies have been responsible for the introduction of coaches of larger dimensions—some upwards of 35 tons in weight : all of these developments have enormously added to the demands made upon the locomotive. A Great Western express fifty years ago weighed no more than 60 tons behind the tender, whereas a modern express weighs anything from 200 to 350 tons, to say nothing of the tender itself which, when no water pick-up apparatus is used, weighs from 45 tons upwards instead of 15 to 25 tons. The following table and Fig. 3 shew the gradual increase of locomotive dimensions.

But designers in their endeavours to solve the problem were face to face with a grave difficulty, namely, the limited dimensions imposed by the strictly limited loading and structure gauge within which any improvement could be made.

In this respect British engineers have been at a great disadvantage as compared with some of their foreign and American rivals, and many of them have more than once regretted that the 'battle of the gauges' was to the 4 ft. 8½ in. party, and have, in spite of its many attendant drawbacks, sighed for the old broad gauge dimensions which would have permitted largely increased boiler and cylinder dimensions and more room for the 'motion' and lengthened bearings. Or, to state the case in other words, with the standard gauge the difficulties are accentuated by the fact that the space allowed outside the rails is less in proportion than for narrow-gauge engines.

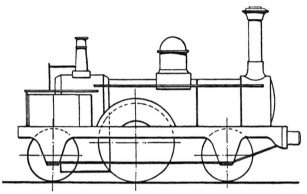

Fig. 1. The *Jenny Lind*.

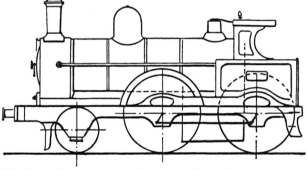

Fig. 2. A *Precedent* engine. London and North Western Railway.

Table shewing the gradual increase of locomotive dimensions

Locomotive	Cylinders Dia. Stroke Inches	Driving Wheels Ft. In.	Adhesion Weight Tons Cwt.	Boiler Pressure Lbs.	Boiler Total Heating Surface Sq. Ft.	Boiler Grate Area Sq. Ft.	Tractive Force at 80% Boiler Pressure Lbs.
"Locomotion" (1825)	10 × 24	4 0	6 10	25	—	—	1000
"Jenny Lind" (1845)	15 × 20	6 0	—	120	780	—	6000
"Lady of the Lake" (1862)	16 × 24	7 6	11 10	—	1086·3	15·0	10,474
Stirling's "Singles" G.N.R. (1870)	18 × 28	8 1	15 0	140	1165	17·6	8536
"Precedents" L.N.W.R. (1875)	17 × 24	6 6	22 10	120	1083·5	17·1	8536
"Experiments" L.N.W.R. (1905)	19 × 26	6 3	48 10	175	2041	25	17,520
"Gt Bear" G.W.R. (1908)	15 × 26 (Four)	6 8½	60 0	225	3400·8	41·79	26,160

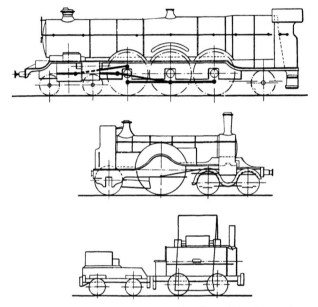

Fig. 3.   The *Great Bear*; a Stirling Single, and the *Locomotion*
shewn to the same scale.

Another feature of recent years has been the
demand for long distance non-stop trains travelling
over one hundred miles without a stop, as for
example Euston-Crewe, Paddington-Plymouth, St
Pancras-Shipley : this has also taxed the ingenuity
of locomotive engineers.

To meet this combination of demands, the past decade has seen a surprisingly large number and variety of new types culminating in such giants as the *White Bear* on the Great Western and the *Baltic* class on the Northern of France.

At the same time this production has been accompanied by a steady tendency towards specialization of duty, the conditions gradually bringing the locomotives of different companies more and more in accord where the work to be done is similar.

Coincident with this feature of similarity in arrangement, we notice that the heating surfaces of boilers have been extended, grate areas have been proportionately enlarged and steam pressures have been raised by as much as fifty per cent. One point worthy of comment, however, is that practically no increase of cylinder dimensions has been made. Locomotive engineers are by no means agreed that increased cylinder dimensions are desirable and, as explained above, there are practical objections to the construction of locomotives with cylinders of exceptional diameter ; hence endeavours in the direction of greater capacity have been confined to the multiplication of the cylinders themselves. Four-cylinder locomotives have been tried on the London and South Western and Great Western, and the Lancashire and Yorkshire railways, and the latest express engine on the North Eastern Railway has three, an arrangement which has also been adopted on the Midland and Great Central. The multiplication of cylinders is not here referred to in connection with the compound system in which four cylinders exist in a very large number of locomotives in Europe and America. Usually there are two inside cylinders and two outside ; the low-pressure cylinders are generally outside and the high-pressure cylinders inside the frames. An interesting exception is provided by the Italian State Railway, in which the two high-pressure cylinders are both on the left-hand side and the two low-pressure cylinders both on the right. A general examination of the highly interesting compound system, which has been looked to by many as a means of securing a higher efficiency is however reserved for treatment in a separate chapter.

In the modern engine the number of driving axles is usually two or three and a leading bogie is the general rule. The Great Western Railway has however a large number with a leading carrying axle.

Bavaria first and now France have adopted a leading and trailing bogie on the same engine.

The gross weight and adhesive weights have reached high figures, amounting in the first case to ninety tons, and in the second, with locomotives with two driving axles, forty, and where three are used, forty-five tons. In exceptional cases weights exceeding eighteen tons per axle are found.

Reference may be made to the devices used in the Continental engines for reducing air resistance, in which the front of the smoke box and the cab front are made wedge-shaped ; also to the means recently adopted to increase the useful effect of the engine, such as superheating, feed-water heating and thermal storage, which however receive special consideration in a later chapter.

In the result British locomotive engineers have dealt with the problem presented to them with conspicuous success, and it is abundantly evident that from the splendid performances of which the details are constantly being made public in railway literature, that high rates of speed with maximum loads are attainable whatever be the form of engine adopted, provided that due regard is paid to the essentials of scientific design and construction.

CHAPTER I

STEAM GENERATION. THE BOILER

It is not too much to say that the success of an engine depends entirely on the boiler, since the power developed is limited by the amount of steam it is capable of supplying. The constantly increasing

demands for more powerful engines have led to a
corresponding development of boiler dimensions to
the extent that the limits imposed by the loading
gauge have almost been reached. Important re-
strictions hamper the locomotive designer in that
the length of boiler is governed almost entirely by
the wheel base, and the diameter by its relation to
that of the driving wheels ; further a large quantity
of high-pressure steam is necessary, which means
that the volume of water to be evaporated in a
given time is also considerable. Limitations of grate
area involve intense combustion induced by forced
draught, and a large heating surface must be provided to utilize satisfactorily the heat thus generated.
Thus the problem to be faced is a more difficult one
than that occurring in stationary or marine practice.

Generally speaking, modern practice has not
meant any great departure from the form of boiler
possessed by the earliest locomotives, and to-day
we have for essentials the features possessed by
Stephenson's *Rocket*, which so effectually disposed of
its rivals in the Rainhill trials by reason of its quick
steam production. For its success it depended upon
the adoption of the multitubular principle, the idea
of which originated, not with the Stephensons, but
with Mr Booth, secretary of the Liverpool and Manchester Railway, and with Séguin in France. This
use of a large number of tubes—usually from 200

to 250—forms the essential feature of the modern boiler and is the chief means of enabling the conditions mentioned above to be carried out.

Four elements compose the locomotive boiler; the inner and outer fire-boxes, smoke-box, and a cylindrical body containing the tubes, which latter run from the inner fire-box to the smoke-box. The heated gases from the fire travel along the tubes to the smoke-box and communicate heat to the water which surrounds both the inner fire-box and tubes. It will thus be seen that a large heating surface is obtained by employing a considerable number of tubes.

Referring to Fig. 4, A is the inner fire-box. It is roughly rectangular in shape, the sides and crown are rolled from one sheet of metal, usually about $\frac{9}{16}$ in. thick throughout. This is riveted to the front tube-plate B and the back plate C, which are both flanged on three sides for this purpose. The metal generally employed is copper, because it does not readily oxidise and resists the action of the fire better than steel. The latter metal is now much used because it is less expensive, and has the same coefficient of expansion as the material of which the rest of the boiler is composed. The bottom of the inner fire-box is formed by the grate, consisting of a number of wrought iron or steel firebars, K, through the spaces between which

the air necessary for combustion finds its way to the
fire.   The firebars are usually $\frac{3}{4}$ in. thick on their
upper surface and from 4 to $4\frac{1}{2}$ ins. deep, tapering

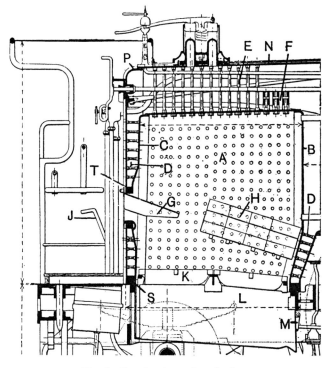

Fig. 4.   Details of a modern fire-box.

off to $\frac{3}{8}$ in. at the bottom. Thickening pieces at
their ends and centres keep them the necessary
distance apart. Under the fire-box is the ashpan, L,
provided with dampers, M, in front and behind, for
regulating the admission of air. Ashpans are of
ample dimensions to prevent accumulated ashes from
interfering with the air-supply, and generally the
bottom is made to contain water for quenching the
ashes. Control of the damper doors is by means of a
handle, J, fixed on the fireman's side of the footplate.

Inside the fire-box above the grate and just
below the bottom tubes is a firebrick arch, H,
which, becoming intensely hot, assists combustion
and directs the hot gases, so that they impinge on
the sides and crown of the fire-box. In this it is
assisted by the deflection plate, G, fitted opposite
above the fire-hole or on the fire-door.

The tube-plate B is drilled to receive the tubes.
For this reason it is made of thicker plate than the
crown and sides, i.e. up to 1 in., or, with steel as
the material, it would be about half this thickness.
To increase its resistance to the action of the fire
an alloy of nickel and copper has been tried, also
a combination of copper for the lower portion and
steel for the upper part which receives the tubes.

The outer fire-box or shell, N, is composed
usually of three steel plates from $\frac{1}{2}$ in. to $\frac{9}{16}$ in.
thick, a wrapper forming the sides and top, the

throat plate in front which receives the boiler barrel,
and the back plate *P* containing the fire-door. The
outer and inner fire-boxes are strongly connected
at the bottom by a foundation ring *S*, the ring round
the fire-hole *T*, and a large number of staybolts *D*.
The latter are very highly stressed, owing to the
enormous pressure acting on the inner and outer
boxes tending to thrust them apart. The magnitude
of this pressure will be realised when it is stated that
it ranges from 200 lbs. per square inch upwards, which
means, in a fire-box of average dimensions, a total
pressure of over 250 tons on the crown and about
400 tons on each side. Fractures of the stays are
frequent enough and are due principally to the
bending action set up by the unequal expansion of
the inner and outer fire-boxes. A double influence
produces this effect, (1) the metal of the inner fire-
box as we have seen has a coefficient of expansion
different from that of the outer ; (2) the inner fire-
box reaches a higher temperature.

The question as to the best material to use for
staybolts is one of the problems of the day. Copper
is in general use, wrought iron and steel less so ; and
recently phosphor bronze and a manganese-copper
alloy have been tried with good results. The stays
are usually from $\frac{7}{8}$ to 1 in. in diameter and pitched
4 ins. apart, so that each supports a plate area of
16 sq. ins. at each end.

Staying the crown of the fire-box is of importance. Two methods are in use, one employing direct stays E and sling stays F (Fig. 4), and the other girder stays. The latter is the more usual. Each has its adherents, those favouring girder stays claiming that their use permits a greater amount of freedom for expansion than the first mentioned. One design of girder stay is shewn in Fig. 5. The girders are of I section and support the roof, their ends taking a bearing on the edges of the inner fire-box as shewn, the stresses being transmitted by the vertical plates to the foundation ring. The roof bars and fire-box crown are connected by slings. These tend to slacken when the box expands on first heating, but as the pressure rises they are put in tension.

The barrel or cylindrical shell, although carrying an enormous pressure, does not present the same difficulty in arranging for the resistance to stress as do the flat surfaces of the fire-box. Stresses in a cylindrical shell are easily calculated, as is the strength of riveted joints, and it is thus simply a question of providing plates of suitable tensile strength and joints equally strong against failure.

The barrel is made up of two or more, often three, rings rolled from suitable plates about $\frac{9}{16}$ in. thick, and either riveted together by straps or hoops, or telescoped, each ring being pushed into its neighbour, as shewn in Fig. 5. The latter is a

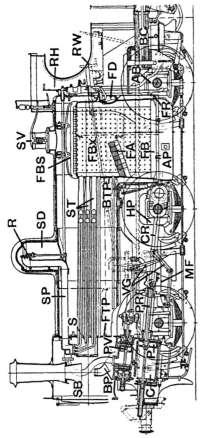

Fig. 5. General arrangement of a modern locomotive.

AB	Axle-box.	FBX	Inner fire-box.	RH	Regulator handle.
AP	Ash-pan.	FD	Fire door.	RW	Reversing wheel.
BC	Brake cylinder.	FR	Foundation ring.	S	Superheater.
BP	Blast pipe.	FTP	Front tube plate.	SB	Smoke-box.
BTP	Back tube plate.	HP	Hornplate.	SD	Steam dome.
C	Cylinder.	MF	Main frame.	SP	Steam pipe.
CR	Connecting-rod.	P	Piston.	ST	Smoke tubes.
FA	Fire arch.	PR	Piston rod.	SV	Safety valve.
FB	Firebars.	PV	Piston valve.	VG	Valve gear.
FBS	Fire-box girder stay.	R	Regulator.		

favoured method, since the position of the largest ring,
next the fire-box, allows a liberal water space and
gives a better provision for circulation, both of which
are rendered necessary by the higher temperatures
incidental to the higher pressures now employed.   In
addition to the staying power afforded by the tubes,
longitudinal stays of 1¼ in. diameter, iron or steel,
run from the smoke-box tube plate, which is from
⅝ to ¾ in. thick, to the back plate of the fire-box.

The junction of the barrel and the fire-box is one
of the weakest parts of the boiler.   Several methods
are in vogue to secure safe connection ; these however need not detain us.

The products of combustion as we have seen are
conducted to the smoke-box and chimney by the fire
or flue tubes, which represent the largest portion of
the heating surface and absorb a large quantity of
heat ; in fact the difference in temperature at the
back and front ends of the tubes reaches from 600
to 700° F.   The tubes generally employed in this
country are of copper, with smooth interior and
exterior surfaces.   They vary in outside diameter
from 1⅝ in. up to 2 ins., and are usually 13 Standard
Wire Gauge thick throughout.   In Europe and
America, however, iron and more recently soft steel
tubes have been introduced.   Copper has the advantage over steel in its more effective resistance
to corrosion due to hard and bad-quality water ;

on the other hand copper tubes are more expensive.
French engineers maintain that with water of good
quality steel tubes are preferable, but admit that
in other circumstances the question is doubtful. In
some cases steel tubes reinforced with copper "safe
ends" at the fire-box tube plate are used.

To increase the available heating surface, tubes
known as Serve tubes, which are 2 ins. or more in
diameter, have been employed to some extent in
recent years.

Fig. 6.   Cross section through a Serve tube.

A section through such a tube is shewn in Fig. 6.
It will be seen that they have longitudinal internal
ribs or fins, which add some 90 per cent. to the
internal heating surface. Their superiority from the
point of view of vaporization has, however, been
contested. The ribs also obstruct the free passage of
soot and cinders, and by reason of their rigidity such
tubes set up severe stresses in the tube plates. The
tubes are slightly inclined from front to back and are
arranged in vertical columns, which facilitates the
dislodgment of the steam from their surfaces.

The method of setting tubes in the tube plate is of the greatest importance, as defective work and vibration lead to leakage. This is also caused sooner or later by the wearing of the tube ends by the abrasive action of cinders, deposits of scale, and temperature variations. The joint (Fig. 7) is made by expanding the tube in the hole, beading over the end and then driving in a ferule of Swedish high-carbon steel. On the Northern of France Railway,

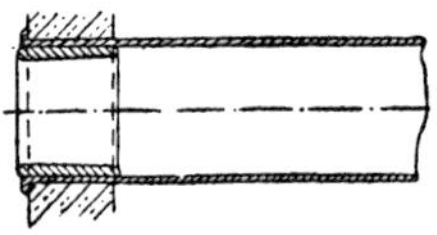

Fig. 7. Details of a joint between tube plate and tube.

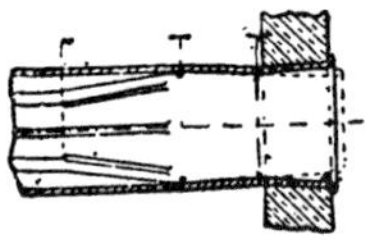

Fig. 8. Method of jointing a tube employed on the Northern of France Railway.

which uses steel tubes, the tubes are swaged to a smaller diameter, rolled and beaded into the copper tube plate (Fig. 8).

The tubes pass into the back plate of the smoke-box, which forms the 'front end' of the boiler. The front plate is fitted with a large dished circular door, accurately fitted to close up air-tight. It is hinged at the side and fastened with a central handle, which actuates a number of dogs pitched equally round the periphery of the door and locked by a second handle on the same spindle. The door permits

access to the tubes so that they may be swept or
'run,' it also enables the accumulated ashes in the
smoke-box to be removed. Formerly and even
to-day on the London and South Western and other
railways, the smoke-box was quite short (Fig. 9),

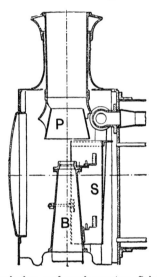

Fig. 9. Short smoke-box, and spark arrester ; Caledonian Railway.

but more recently, following American practice, it
has been much lengthened to form a cylindrical
projection from the boiler known as the 'extended'
smoke-box (Fig. 10). The means for obtaining

forced draught, namely the blast pipe P, is situated
herein, as is also a part of the superheating apparatus,
S, where such is employed ; further the steam pipe E

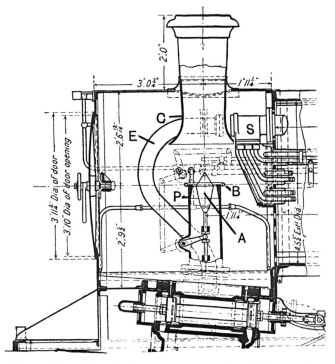

Fig. 10. Extended smoke-box, variable blast orifice and piston valve.
London and North Western Railway.

from the boiler to the steam chest passes through it, itself acting to some extent as a superheater.

The primary function of the smoke-box and its equipment is a most important one, namely the production of a partial vacuum and hence a draught, upon which depends the economical burning of the fuel and at such a rate of combustion as is necessary for satisfactory steaming. These qualifications depend largely upon the disposition of the blast-pipe orifice in its relation vertically to the chimney top, together with a correct height from the boiler centre line and a correctly proportioned chimney.

The action of the blast pipe as at present understood is as follows : the exhaust steam escaping from the cylinders issues through the contracted orifice of the blast-pipe at a high velocity and expels part of the contents of the smoke-box, thereby creating a partial vacuum. To destroy this vacuum the products of combustion are forced through the tubes by the atmospheric pressure outside the ashpan and fire-hole, thereby creating a fierce draught on the fire. The amount of draught is measured by the vacuum in the smoke-box, which may be anything up to 7 or 8 inches of water.

Another and important function of the smoke-box and its equipment is to deal with the enormous and variable quantity of air needed for combustion which is drawn in at the fire-box. This air, it must be remembered, is immediately expanded six to eight times by rise of temperature, and on arrival at the smoke-box occupies from two to three times its original volume, the difference being accounted for by the rapid cooling which takes place in its passage through the tubes.

Mr Hughes, chief mechanical engineer of the Lancashire and Yorkshire Railway, carried out a year or two back some interesting experiments with a view of ascertaining the value of long and short smoke-boxes. A higher vacuum was obtained in every case with the extended smoke-box which, as Mr Hughes points out, tends to prove that the long box serves as a reservoir, thus assisting the maintenance of draught between each exhaust, and so modifying the intermittent character of the blast. This was also verified by the action in the U-shaped glass tubes or manometers partially filled with coloured water used to observe the vacuum. With the extended smoke-box the water remained quite steady, and only moved when the steam-discharge up the chimney was altered ; whereas with the short box the water was in a constant state of agitation, rising and falling with each exhaust. Further, the steam pressure was better maintained in the extended smoke-box engine.

The shape of the chimney has also an important bearing on the character of the blast. This has formed the subject of experimental determination at the Purdue University, U.S.A., by Professor Goss following on those conducted in 1896 on the Continent by Von Borries. The outcome of these experiments was that it is preferable to use a chimney of conical shape, that the diameter should be small enough to

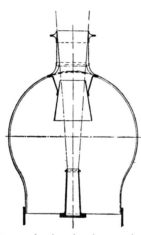

Fig. 11.   Diagram shewing the piston action of the blast.

cause the cone of exhaust steam to strike the barrel of the chimney, and of such a length that the steam would form a fluid piston capable of setting up by its movement a sufficient vacuum in the smoke-box. To effect this, the taper of the chimney should be such that the puff of steam continues to fit until

it finally emerges into the atmosphere (Fig. 11).
Practice has shewn, particularly in the case of certain
British locomotives of recent construction, that a
reduction in the length of chimney does not inter-
fere sensibly with the blast. Professor Goss has
shown that the practice of prolonging the chimney
on the interior side of the smoke-box below the
petticoat pipe, a practice in vogue on several American and European lines, has caused a diminution in
the effect of the blast, and it is therefore preferable
to abandon it.

Recent practice, as we shall see in another chapter,
has adopted a blast pipe with a variable nozzle
(A, Fig. 10), the orifice of which can be adjusted to
the demand for steam. American engineers have
long since recognized the theoretical superiority of
this arrangement, and its adoption has also found
favour in Europe. Some difficulty appears to have
been found in maintaining the parts in working order,
but this is not a serious matter.

The smoke-box contains also the spark arrester
(S, Fig. 9), by which the throwing of live coal is
diminished, and the dead cinders are kept in such
a position in the smoke-box as to remain clear of
the bottom rows of tubes. The pattern used by
Mr J. F. McIntosh on the Caledonian Railway,
consists of a V-plate interposed between the smoke-
box tube plate and the blast-pipe, and extends from

the bottom of the smoke-box to a level just above the top row of tubes. This deflector plate is pivoted on brackets at the back of the blast-pipe, thus allowing the plate to be swung round and the tubes to be cleaned.

In working, the ashes drawn through the tubes are deflected away from the strong current caused by the blast, and instead of settling down gradually and blocking the lower tubes, the cinders roll back to within the V of the deflector, being thus kept away from the tubes.

The cylindrical shell of the boiler is lagged, i.e. covered with some non-conducting material to prevent loss of heat by radiation and to impart a finished appearance. Wood and felt with an outer covering of Russian iron were extensively used, but when running the wood would frequently catch fire, necessitating the stoppage of the engine to extinguish it.

The writer once had a very unpleasant footplate experience on the Great Central Railway. On entering the Woodhead tunnel, which is one of the longest in the country and only of sufficient width to accommodate one train, the lagging fired with the formation of clouds of dense, suffocating smoke, which enveloped the footplate and engine. In the confined space of the tunnel the conditions are better imagined than described, especially as it was impossible to stop until the open air was again reached.

Modern engines are fitted with asbestos coverings and an outer envelope of sheet steel.

Boiler fittings and accessories, with the exception of the injector, which will be dealt with elsewhere, do not concern us, since they play no part in steam raising.

Having now obtained a general idea of the construction of the boiler, we shall in a succeeding chapter examine some of the modern developments introduced to improve its working. To appreciate the significance of these, some attention must now be given to combustion and vaporization.

CHAPTER II

COMBUSTION AND VAPORIZATION

FUELS depend for their heating value upon the presence of such calorific constitutents as carbon, hydrogen, and compounds of those bodies called hydrocarbons. The chemical union of these with oxygen forms the familiar process of combustion. It is accompanied by the evolution of light and heat, which latter is transferred to the water in the boiler in the way we have seen.

The principal chemical combinations resulting

from the combustion of hydrogen, carbon, oxygen, are carbon monoxide (CO) and carbon dioxide or carbonic acid gas (CO_2). The former results from the partial combustion of carbon with a limited supply of oxygen, the latter by perfect combustion secured by a copious supply of oxygen. Two important hydrocarbons are also formed in gaseous form, namely, methane or marsh gas (CH_4) and ethylene or olefiant gas (CH_2). In the locomotive the necessary oxygen is obtained from atmospheric air which contains oxygen and nitrogen in the proportion, roughly, of one part of the former to four parts of the latter by volume, or 56 parts of nitrogen to 16 parts of oxygen by weight. Thus, to obtain one cubic foot of oxygen, it is necessary to supply five cubic feet of air; for one pound of oxygen, we must have $\dfrac{16+56}{16} = 4\frac{1}{2}$ lbs. of air roughly. Nitrogen, for the purposes of combustion, is inert. It becomes heated, however, in the process and so absorbs a considerable proportion of the heat developed in the fire-box, thus limiting the maximum temperature obtained. The heat evolved, or the heating value, depends upon the relative proportions of the constituents and has been determined for the separate elements as follows: one lb. of hydrogen when burnt with oxygen to form water evolves 62,032 British Thermal Units (usually expressed B.T.U.), which suffices to evaporate 64·2 lbs.

of water from and at 212° F.*   It requires for its
combustion 8 lbs. of oxygen.   One pound of carbon
when completely burnt evolves 14,500 B.T.U. which is
sufficient to evaporate 15 lbs. of water from and at
212° F. For combustion $2\frac{2}{3}$ lbs. of oxygen are required.
When partially burned, and carbon monoxide is
formed, only 4400 heat units are evolved capable of
evaporating 4·55 lbs. of water from and at 212° F.
For its combustion $1\frac{3}{4}$ lbs. of oxygen are required.

The fuel in most common use for locomotives is
coal.   Mineral oil is used to some extent on the Gt
Eastern Railway, and more extensively in Russia and
America. Coke was formerly employed exclusively
owing to its smokeless combustion, and it was not
until the invention of the brick fire-arch that coal-
burning was rendered possible.

The principal varieties of coal found in this country
are anthracite, semi-anthracite, and semi-bituminous.
The two last mentioned are used chiefly for steam
raising, the best being the Welsh steam coals, which
burn readily without the formation of black smoke.

---

* It is usual in expressing evaporation results to use a common
standard, namely, that of the number of pounds of water at a temperature of 212° F., which would be converted into steam of the same
temperature by the application of the same number of heat units.
Each pound of water so evaporated would take up 966 B.T.U., hence
we get equivalent evaporation $= \dfrac{\text{heating value in B.T.U. per lb.}}{966}$.

The following table gives the heating values of the principal varieties of fuel.

Fuel	Heat of Combustion in Thermal Units per lb.
Best Welsh Coal	15,000—16,000
Newcastle	14,820
Derbyshire and Yorkshire	13,860
Lancashire	13,900
Scotch	14,100
Coke	12,500
Mineral Oil	19,000—19,500

These are theoretical values calculated from the chemical composition, and upon the assumption that one pound of pure carbon is capable of evaporating 15 lbs. of water from 212° F. But, as we shall presently see, the practical results differ considerably from the theoretical. Coal, however, is not usually bought by railway companies on a basis of chemical constituents, although tests for heating values are regularly made for checking purposes. These values may be calculated from the composition as found by chemical analysis or determined by means of a calorimeter, the latter method giving perhaps the more certain results. This is not the place to examine the phenomena of thermal changes which accompany chemical reactions, but in estimating calorific values they are of importance. For, just as chemical union may be

A. L.   3

accompanied by the evolution of heat, so a corresponding dissociation requires its expenditure and
must be taken into account. Again, the constituents
may combine to form compounds other than the
products of combustion as, for example, where a fuel
contains both oxygen and hydrogen, some of the
hydrogen will combine with some of the oxygen to
form water, consequently no heat will be available
from this hydrogen. Thus the weight of hydrogen
available in 1 lb. of coal for calculating the heating
value will be given by $H - \dfrac{O}{8}$, since 8 parts by weight
of oxygen unite with one by weight of hydrogen to
form water.

Determination of the calorific value by the calorimeter may be made by a Mahler apparatus, which
consists of a steel shell containing a pound of combustible. This is ignited by an electric spark and
is burnt instantaneously by the aid of pure oxygen
introduced under a pressure of 400 lbs. per square
inch. Before ignition, the shell is immersed in a
water calorimeter.

We have said that the practical heating power
of coal differs from its theoretical calculated value.
This is accounted for, first, by the waste of fuel and,
secondly, the inability of the boiler to utilize all the
heat generated in the fire-box. Taking the last mentioned cause first, the evaporation which takes place

in a boiler is only about 7 to 8 lbs. of water per lb. of coal, representing about 9½ lbs. from and at 212° F. and a boiler efficiency of probably about 66 per cent. This low evaporative duty is chiefly due to the high temperature retained by the gases after they leave the tubes, about 25 per cent. of the available heat being wasted in this manner. Part of this loss is unavoidable, as it is impossible for the gases to cool down to the temperature of the steam, some head of temperature being necessary to enable the heat to penetrate the metal.

Unskilful stoking is also a source of waste, so much so, that it is the practice on most lines to give bonuses for coal saving. Excessive coaling means that sufficient air cannot reach a portion of the fire, hence some of the coal being only warmed will be distilled and part with its valuable and volatile hydro-carbon constituents in the shape of unburnt gas; a proportion of the remainder will be incompletely burnt to carbon monoxide, which means that only 4400 heat units per pound of carbon are generated instead of 14,500.

Much, however, can be done by proper manipulation of the damper and firehole door, and a good fireman will be influenced by his position on the road when firing up. Weather conditions also exert a great influence on the fireman's work, an engine being generally found to steam best against a head wind, and worst with a side wind; in the former case the air is forced well into the ashpan and through the fire bars, whilst in the latter case the wind rushing across underneath the engine has a tendency to suck out the air in the ashpan, acting much as a steam ejector would do.

A considerable quantity of small coal is drawn through the tubes by the fierce draught, and as most of it is in an incandescent state, an appreciable loss occurs. The loss is greatest, of course, when using the small coal commonly known as 'slack,' and with Welsh coal, which splits up into small pieces when heated, instead of caking together in a pasty mass like the bituminous varieties. In recent years spark-throwing has been much diminished, as we have seen, through the introduction of spark arresters. A fierce blast is also unfavourable to coal economy, not only because it tends to increase spark-throwing, but the gases are drawn through the tubes at such a high velocity that they have not time to give up their heat and the smoke-box temperature rises to an excessive degree. Other causes of inefficiency are contraction of the flue-way area of tubes by ferules or otherwise, and the formation of a non-conducting deposit or scale due to the presence in the water of carbonates and sulphates of lime.

The available information concerning the amount of heat lost in the working of the locomotive has recently been added to by the results obtained by the Breslau Royal Railway Department from a definite series of vaporization tests. It is worth while stating the results reached. (1) The heat escaping in the smoke gases was 20 to 23 per cent. of the total heat value of the coal. (2) The loss by the combustible components found in the residue varied, according to the design of the locomotive, from 5 to 11 per cent. of the total heat value of the coal. (3) The loss by radiation, smoke and spark production was about 5 per cent. (4) The efficiency of the boiler therefore varied from 60 to 70 per cent. (5) The average temperature in the smoke-box was from 330° to 380° C. (716° F.). (6) The mean temperature of combustion in the fire-box was found to be 1485° C. (2705° F.). (7) The mean specific heat of the smoke gases of average composition reduced to 0° C. is 0·324 for a smoke-box temperature ranging from 350—400° C. (662—752° F.) and 0·35 for the fire-box and tube temperatures. (8) The gases with the loco-motive in full running and 350° C. (662° F.) in the smoke-box were found to contain an average of 11 per cent. carbon dioxide and 0·6 per cent. carbon monoxide. From one kg. (2·2 lbs.) of coal 11 to 12 cub. ms. (388—423 cub. ft.) of smoke gas were obtained at 0° C.

The inability of the boiler to utilize all the heat generated from the fuel follows from the nature of heat transmission to the water. The heat evolved in the fire-box is propagated in two ways: by direct conduction and by radiation. By the first the heat is usually considered as propagated by the hotter molecules heating the neighbouring colder molecules of the plates; by the second the transference takes place without the intervention of matter by etheric waves set up by the vibration of the heated molecules. It is in this manner that radiant heat (and light) reaches the earth from the sun.

If there were no radiant heat the temperature of the fire-box gases would be extremely high, and all the heat evolved would be available for heating them. As great a difference of temperature as 1800° C. (3272° F.) can exist between the calculated and ascertained temperatures in the fire-box, which affords a measure of the losses due to the radiant heat.

The rate of conductivity of metal is such that a few degrees' difference in temperature on each side of a plate is sufficient to account for the transmission of a large quantity of heat. Thus it has been found that a vaporization of 200 kgs. per sq. m. (41 lbs. per sq. ft.) per hour in a copper fire-box corresponds to a difference of only 4·7° C. (40·4° F.) between the two faces of a plate 13 mm. ($\frac{13}{25}$ in.) thick. M. Nadal, locomotive engineer of the French State Railways, has stated that in steel tubes of 2·5 mm. ($\frac{1}{10}$ in.) thickness a vaporization of 60 kgs. per sq. m. per hour (12·3 lbs. per sq. ft.) means a corresponding difference in temperature of only 1·7° C. (35° F.).

From the hot gases to the fire-box walls, and from the latter to the water, heat is transmitted by external conduction. The coefficient of conductivity between the plates and the water is high, and the temperature difference small; in fact the temperature at the surface of the plate is at the most 15—20° C. (59—68° F.) higher than that of the water. On the contrary, the coefficient of conductivity between the gases and the plates is small, which necessitates increasing the contact surface to the greatest possible extent. Hence we get the Serve type of tube. It is owing to the high conductivity of the metal and the high coefficient of exterior conductivity between the metal and water that the fire-box surfaces can readily absorb the radiant heat.

It follows, therefore, that the direct heating surface obtained from the fire-box and portions of the adjacent tube produces a more effective evaporation than the rest of the flue area and accounts for from one-third to one-half of the total quantity of steam generated.

The power developed by the boiler is therefore in reality limited by the grate area and the maximum amount of coal consumption. With a given grate area and a given strength of draught, a definite quantity of coal can be burnt per hour. A large

grate area means that a large amount of heating
surface must be provided to ensure the efficient utili-
zation of the heat. Thus the evaporative power of
the boiler depends upon the ratio of heating surface
(HS) to grate area (GA) and rate of coal consumption.
It is generally stated in pounds of water evaporated
from feed temperature per square foot of heating
surface per hour.

The ratio $\frac{HS}{GA}$ varies between 60 and 100, the
average being about 80. With a ratio of less than 60
the flue area will be reduced to require a very sharp
blast which we have seen to be a disadvantage, while
increasing it to over 100 means crowding of tubes
or undue elongation of them. The former obstructs
the water circulation, and the latter means increased
frictional resistance to the flow of the hot gases and
much reduced temperature in the last foot or two of
length, which becomes then of little value as heating
surface.

The rate of coal consumption reaches 150 lbs. or
more per square foot of grate area per hour for short
periods and fast running ; it may average on a run
with a train load of, say, 300 tons, 90 lbs. per sq. ft.
per hour.

The amount of water evaporated averages 30 lbs.
per indicated horse-power per hour, measured from
the tender, of which at least 21 lbs. are required

for the engine, the balance being consumed for
working the injector, blower, brakes, and by blowing
off at the safety valve. Or, it may be stated that as
much as 13 lbs. of water can be evaporated from one
sq. ft. of heating surface per hour. An average of
3 sq. feet of total heating surface per indicated horse-
power may be taken as an approximation.

The modern big boiler has another advantage
besides that due to the large grate area and heating
surface, namely, its capacity for carrying a large
volume of hot water. Thus, should the steam pressure shew a tendency to fall when nearing the top of
a long bank, the feed can be shut off, thus temporarily
increasing the boiler power by some 25 per cent.,
owing to the fact that the latent heat of evaporation
only has to be supplied. Some three or four miles
can be run with the feed shut off without letting the
water-level drop dangerously low.

CHAPTER III

INCREASING THE USEFUL EFFECT OF THE BOILER

WITH the object of increasing the efficiency of the
standard boiler, and of obtaining increased power
from it, numerous devices and experiments have been
tried in recent years, some of which have yielded

sufficiently satisfactory results to justify their permanent adoption. Some of these methods such as
coning the boiler, varying the fire-box contour, the
employment of water tubes, pre-heating the feed
water, thermal storage and superheating, will now be
examined.

Cone Boiler. This type of boiler, an example of
which is seen in the *Gt Bear*, Fig. 3, forms a prominent
feature in recent locomotives of the Gt Western
Railway, designed by Mr Churchward. Some of the
boiler rings, instead of being truly cylindrical, form
the frustum of a cone, with the result that the
largest cross-sectional area of the boiler barrel is
in the locality where we have seen the highest in-
tensity of combustion takes place, consequently where
the heating surface is the most valuable, namely, close
to the smoke-box. This is a distinct and definite
advantage, since it provides a greater area of water
line, an increased steam capacity and, by the larger
diameter being arranged to coincide with the line of
the fire-box tube plate, much more water space at
the sides of the tubes.

It has also materially contributed to the reduction
of priming or foaming and it enabled the dome, always
a source of weakness, to be dispensed with and at
the same time to secure dry steam. This important
result has also been obtained by the employment of a
fire-box with a flat top, the most conspicuous example

of which is the invention of a Belgian engineer, M. Belpaire.

Belpaire Fire-Box (Fig. 12).   In this the outer

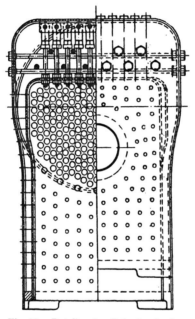

Fig. 12.   Details of a Belpaire fire-box.

wrapper is made flat on top, parallel to the roof of the inner fire-box, to which it is tied by stays similar to those used for connecting the sides of the ordinary

fire-box. The first two rows only are provided with
sling stays for securing vertical flexibility to the
front of the box, and cross-stays connect the sides
of the walls of the outer wrapper above the inner
box. With the flat top the area of the water line at
the hottest part of the boiler is increased and more
steam space provided, since the girder stays, which
take up a large portion of the heating surface of
the top of the ordinary box, and are thought by
some to hinder the free rising of the steam bubbles,
are dispensed with. Mr Churchward states that less
trouble has been experienced on the Gt Western
Railway with the Belpaire box than with the round
top. It is therefore not surprising to find that it is
increasing in favour, and to-day it is employed by the
Gt Central, Midland, North British, Lancashire and
Yorkshire, Gt Eastern, the last to adopt it being
the London and North Western.

Wootten Fire-Box (Fig. 13). The deep, round-
topped fire-box spreading wide outside the frames—
a feature of Mr Ivatt's famous *Atlantic* engines on
the Gt Northern Railway—is known as the Wootten
fire-box. Its employment is possible only when the
rear pair of coupled wheels are set well forward and
a pair of trailing wheels used to carry it. Hence
it is in favour on the 4–4–2 and 4–6 -2 types (see
p. 123). The lateral walls are strongly inclined towards
the exterior, for example with a barrel of 5 ft. 8 in.

diameter, the width of the base of the fire-box reaches as much as 7 ft. 2 in. The Wootten box originated in America, where it is still largely used. Designed, however, to burn inferior fuel, the recognized advantage of employing high class coal has led to a marked development with the object of reducing

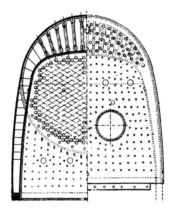

Fig. 13.   Details of a Wootten fire-box.

its solidity and cost of upkeep. Moreover, the inclination of the lateral walls, if carried too far, retards the ascending currents of steam since the course of circulation in a boiler is upwards in the fire-box, and downwards in the smoke-box end.

American Fire-Box. Fig. 14 represents one of
the most recent types of large American fire-boxes.
It is simple in form, with straight lateral walls and
sloped back plate. The lower part of the front tube
plate is also sloped, which allows the box proper

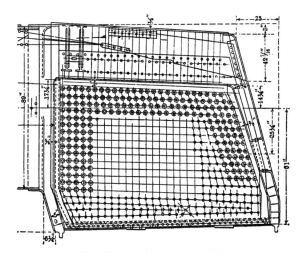

Fig. 14.   American type fire-box.

to be prolonged into the barrel for about 3 ft., so
constituting a combustion chamber increasing to a
certain extent the evaporative efficiency.   American
fire-boxes constructed without spread of the lateral
walls are called 'wagon top.'

Jacobs-Schupert Fire-Box (Fig. 15). In this, the latest American development, the usual arrangement of flat plates supported by stay bolts has been abandoned, except in the front and back sheets. Side and wrapper plates have been replaced by channel-shaped sections riveted together. These are stayed by stay sheets interposed between the sections. All seams are submerged, and no joints are exposed to the direct currents of heat and gases. Owing to the

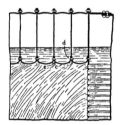

Fig. 15.   Jacobs-Schupert fire-box.

irregular outline thus formed for the crown and sides, the available heating surface of the hottest section of the boiler is enlarged without increasing the size of the grate area, and the arched concave construction of the sections ensures that there will be no undue local stresses, the shape of each section being such that it will expand or contract with variations in temperature and produce only small stresses on adjacent sections.

Stayless Boiler. A number of attempts have
been made in Europe to dispense altogether with
the fire-box. Herr Lenz in Germany some years
ago introduced a corrugated form of fire-box which
was claimed to be sufficient in itself to support the
tube plates, and no further stays were used. After
some disastrous explosions, however, they were withdrawn from service. Vanderbilt, on the Prussian
Railway, has also employed a stayless boiler with a
corrugated fire-box.

Water-tube Boilers and Fire-Boxes. In spite of
the extensive adoption, within recent years, of the
water-tube type of boiler for both land and marine
service, little has been heard of the possibilities of
this type of steam generator in connection with the
railway locomotive. This is more particularly striking
in view of the quick steaming requirements of the
modern locomotive and the special qualities which
appear to be possessed by the water-tube boiler for
meeting them.

The apparent diffidence with which the water-
tube boiler problem has been treated by locomotive
engineers has, however, met with some notable
exceptions in the case of Herr Brotan, the cele-
brated Austrian engineer, and Mr D. Drummond,
the Locomotive Superintendent of the London and
South Western Railway, who for some years past have
consistently made use of water-tubes with a great

amount of success. The feature of the Brotan system (Fig. 16) is the replacement of the ordinary inside fire-box by a system of water-tubes, involving the elimination of the customary water space and stays. The boiler proper is divided into two cylindrical barrels fixed parallel to each other, the lower and main portion containing a number of fire-tubes. Connection with the upper barrel is made by means

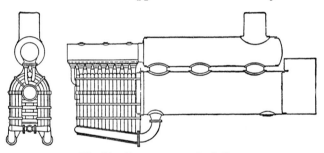

Fig. 16.   Brotan water-tube boiler.

of two necks. The upper ends of the water-tubes, composing the fire-box, are fixed in the rear end of the upper barrel, which is therefore of thicker plate. The lower ends of the water-tubes are expanded into a rectangular shaped water-circulating chamber of steel, connection between which and the back end of the main barrel is made by a large pipe. Facilities for inspecting and cleaning the water-tubes are provided by means of a number of removable doors on the underside of the circulating chamber, giving access to corresponding holes into which the water-tubes themselves are expanded. The fire-box tubes are encased in fire-clay and a covering of sheet steel plates. One example of this type of boiler, constructed by Messrs Beyer, Peacock & Co., exists at the works of the Mannesman Tube Company. On the Austrian State Railways, they have been employed with notable success since 1901.

Schneider Boiler. Somewhat similar in type was the boiler of a locomotive shewn at the recent Nancy Exhibition by Messrs Schneider & Cie of Creusot. The boiler consists of an upper drum, containing water and steam, extending along the whole length of the boiler and connected by small diameter tubes to four water collectors. The rear pair of the latter are placed one on each side of the fire-box.

Each pair of front and back collectors are inter-connected by cast steel tubes, and communication between these and the upper drum is made by a return water-tube. The back water-tubes are splayed to enclose the grate which, together with the tubes themselves, forms the inside fire-box. The tubes are interlaced at the top to screen the drum from the direct action of the flames, and the tubes in each of the two outside rows are closely juxtaposed to pre-vent the escape of the hot gases. The front and back portions of the fire-box are built up of fire-brick,

which material is also used as a covering for the
lower portions of the water-tubes and collectors. The
front tubes are disposed so as to form a horizontal
flue for the passage of the products of combustion
to the smoke-box, the tubes being arranged in rows
in a longitudinal direction. As in the case of the
back group, escape of gases from the sides is prevented
by close contact of the tubes in the two outer rows.

The tubes connecting the drum and collectors are
inlet on the underside of the drum, and a very low
level of water suffices to cover them entirely. This
is held to constitute an important advantage peculiar
to this type of boiler, as the volume of free water
comprised within the maximum and minimum levels
is sensibly greater than that which is available within
the same limits in cylindrical boilers of the ordinary
smoke-tube type. This means a greater reserve of
energy, which can be drawn upon when long gradients
have to be negotiated.

The surface of ebullition remains nearly constant
whatever may be the height of the water-level in the
drum, because it is always in the neighbourhood of
the horizontal diameter of the drum. Such is not the
case in boilers of the ordinary type, in which the surface of ebullition diminishes progressively with the
height of water-level, so that priming, which results
from such a diminution of evaporating surface, cannot
take place in the new type of boiler. The outside

4—2

covering is made up of removable segments, prolonged at the front end to form a smoke-box.

Riegel on the Southern Pacific Railroad of America
uses a water-tube boiler on express passenger engines.
The water-tubes are located in the fire-box, and the
foundation ring, which is of cast steel, has water-
pockets cast in it at the sides, beyond the grate and
throughout its length, thus forming lower terminations for two nests of water-tubes. These extend
from the pockets diagonally upwards to the crown
plate, which is slightly depressed to keep the upper
tube terminations flooded. Above the crown plate
is provided a staying cylinder, which, with the crown
plate, makes a double thickness at the crown for
tube ends ; this cylinder has sufficient flexibility to
allow for expansion and contraction. The tubes can
be withdrawn through the water-pockets which are
fitted with removable plates. This fire-box has no
less than 768 sq. feet of heating surface.

Marine Type. The marine type of water-tube
fire-box (Fig. 17) employed with satisfactory results
on the Northern of France Railway deserves mention.
An engine so fitted was shewn at the 1910 Brussels
Exhibition after having covered 33,000 kms. on the
road. In vertical cross-section the fire-box resembles
the Wootten overhanging type, affording accommodation for a group of splayed water-tubes which form
the side walls of the box. The tubes are expanded,

in the manner peculiar to marine practice, into a
header or cylinder at the top of the box and, at
the bottom, into two water legs or drums extending
laterally along the sides of the fire-box. The high
pressure of 255 lbs. per sq. in. is employed.

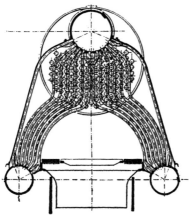

Fig. 17.   Marine type of water-tube fire-box ; Northern of France
Railway.

A fire-box of this type has been fitted to one of
the new *Baltic* 4–6–4 type engines of the Northern
of France Railway, illustrated in Fig. 19.

It is, however, Mr Drummond who has made the
most consistent use of the water-tube principle, and
practically all the engines on the London and South
Western Railway are so fitted.

Fig. 18. 4–6–0 type 4-cylinder simple expansion engine, with water-tube fire-box; London and South Western Railway.

Fig. 19. 4–6–4 (*Baltic*) type 4-cylinder compound express locomotive : Northern of France Railway.

Fig. 20. 4–4–2 (*Atlantic*) type express locomotive ; North British Railway.

They have proved themselves to be more economical in coal consumption than similar engines
fitted solely with flue tubes. This is doubtless due
to the direct cycles of water circulation through the
tubes and about the fire-box in general, which cause

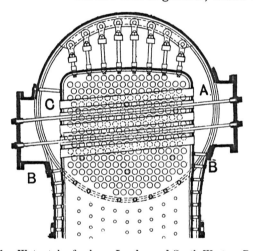

Fig. 21. Water-tube fire-box ; London and South Western Railway.

rapid heat absorption and prevent scale formation.
The writer can testify to the remarkable absence of
scale as a result of an inspection of these engines
immediately after being taken off the road for repairs.
Transverse water-tubes are employed as shewn in
Fig. 21. They are of mild steel, slightly inclined

and rolled into the lateral walls of the inner fire-box. Transverse stays are passed through some of the tubes to stiffen the box suitably. Access to the tubes is obtained by a hinged door at the side, accurately faced to form a steam-tight joint with the faced rectangular castings on the outer fire-box. They form a fine example of hand filing and of a metal-to-metal joint. Incidentally the illustration shews a method of slinging the inner fire-box without the use of girder stays. The sling bolts are in couples with nuts bedding on crosspieces, leaving the nuts free to lift up when the fire-box rises by expansion. The rising pressure, however, brings the nuts back again on their seating.

The success of this well-tested water-tube ar-rangement has led Mr Drummond to pursue his investigations further, for which purpose he built, some time ago, a locomotive entirely on the water-tube principle. The results obtained with this are not yet known.

Water Softening. The incrustation deposited on the walls of the tubes by the use of hard waters, i.e. water containing carbonate and sulphate of lime, and magnesia in solution, is an extremely bad conductor of heat and its presence in any quantity needs not only more heat to evaporate it, but leads to over-heating or burning of the plates. The trouble from this cause increases as the pressure and temperature

of the steam rise. In some cases it has been found
that water which gave little or no trouble at 160 lbs.
pressure was practically unusable at 200 lbs. More
attention is now paid to the treatment of water before
it is used, and large water-softening plants have been
installed in districts where the water is notoriously
hard. The systems used differ in the method of
adding the chemicals, but they depend essentially
upon the principle that free carbon dioxide (CO_2)
assists to keep the lime in solution. If an excess
of lime be now added to the water, the CO_2 is
neutralized and the whole of the lime, including
that originally present in the water, is thrown down
as a precipitate. Soda ash is also employed.

An electrolytic method of treatment has been
experimented with in America which, although costly,
reduced the incrusting solids from 40 grs. to 6 grs.
per gallon.

Briefly stated, the process consists in submerging
aluminium or iron plates in the water and then
passing an electric current through the plates which
are connected up in series. The plates enter into solution in proportion to the quantity of water treated.

Oil-burning Apparatus. Liquid fuel is used on
the Gt Eastern Railway and on the locomotives
running in the oil-field countries such as Southern
Russia, the Far East, and the Southern States of
America and Mexico. On the Southern Pacific

Railroad alone, nearly 1000 locomotives are of
the oil-burning type.   In these engines the oil is
carried in tanks built to fit the coal space in the
tender.

The burner used is of the flat-jet type consisting
of a flat casting, divided longitudinally by a partition
over which the oil flows as it is admitted to the upper
cavity.   The lower cavity receives the steam for the
jet which strikes the oil flowing over the partition,
spraying it into the furnace which has refractory fire-
bricks built in on the lower sides to prevent the oil
blast impinging against the sheets.   The aim is completely to atomize or break up the oil near the burner
tip in order that it may be immediately vaporized.
The steam for atomizing is obtained from the dome.
Other methods obtain atomization with compressed
air which, however, is liable to produce in the furnace
a more intense local heat than is desirable.   With
the steam jet the oil is sprayed and broken up so
as to allow the air admitted through the proper
dampers to mix and the oil to be consumed completely without damage to the plates.

The best evaporative results obtained from steam
jet burners give an evaporation of 13 lbs. of water
from and at 212° F.   With the air-jet burner supplied
with heated air at 5 to 7 lbs. per sq. in., 16 lbs. of
water can be evaporated.   Tests made on oil-burning
locomotives shew that temperatures ranging from

2500 to 2750° F. are obtained with the steam jet type.

On the Gt Eastern Railway, the Holden system is employed whereby creosote is used as an auxiliary to the ordinary fire. An inner steam jet and an outer

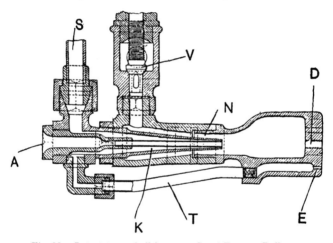

Fig. 22.   Latest type of oil burner ; Great Eastern Railway.

annular jet of oil spray are used, which play over a bed of incandescent fuel.

In the latest form of the Holden steam jet burner, illustrated by Fig. 22, the spray is projected from a series of holes *D* arranged at a slight angle, so that the streams of atomized mixture shall

converge after leaving the mixing chamber. Steam is
projected from a series of holes E, and supplied by
a pipe T from the main supply entering at S; the oil,
which enters at slight pressure along the pipe at
the side, is controlled by a screwdown valve V, in
its passage to the base of the outer cone N, along
which it is drawn by an annular steam jet supplied
at about 60 lbs. pressure to the inner cone K. The
steam jet also induces a jet of air from A by way
of the central tube. The calorific value of the crude
petroleum used varies from one to one and a half
times that of coal. Oil fuel has also been employed
on the engines working through the Arlberg tunnel on
account of the smokeless combustion of the fuel.

Exhaust Steam Injectors. The apparatus most
generally in use for feeding the water into the boiler
is the Giffard injector, the action of which affords one
of the most interesting problems in thermo-dynamics.
It would be impossible within the limits of this
chapter, to examine the theory of its working; it
must suffice to state that it depends upon a rush of
steam from the boiler at an enormous velocity to
induce the flow of a corresponding stream of cold
water, by which the steam is condensed. The velocity
attained by the combined stream of cold water and
condensed steam is sufficient to cause it to enter the
boiler against the same internal pressure as that of
the steam itself. In the diagram Fig. 23 steam enters

at *A* and passes through the nozzle *G*. Water is drawn in at *E* and mixes with the steam in the combining tube *C*, and is carried forward together with the condensed steam with great velocity to the delivery tube *D*, thence into the boiler. The maximum velocity is reached at the narrowest part of the delivery tube. The break at *O* is the overflow to allow the excess of water or steam to escape. It is, as we shall see in the next chapter, highly desirable that the temperature of feed-water should be raised to the highest possible degree at which the injector will work before it enters the boiler; and if this can be accomplished by means of exhaust steam, which would otherwise go to waste, it is easily apparent that a great saving must of necessity be effected.

An injector depending for its working mainly on exhaust steam was introduced some years ago by Messrs Davies & Metcalfe, and recently they have greatly improved the apparatus. Leading off from the blast-pipe of the locomotive is a branch pipe, by means of which steam is conveyed to a grease separator, where the exhaust steam is freed from any oily impurities or water present. The steam then passes to a central exhaust nozzle *S* (Fig. 23), at the mouth of which it comes into contact with the feed-water from *E*. Condensation takes place, and a high velocity is thus imparted to the combined jet, which then flows forward through a draught tube. At the end of this it meets with a second supply of exhaust steam, which imparts a further supply of energy to it, and the combined jet enters the combining nozzle, C, where complete condensation takes place, and its velocity is still further increased. Then it passes to the delivery nozzle, D, where its velocity energy is transformed into pressure energy and so to the boiler, F. The exhaust steam

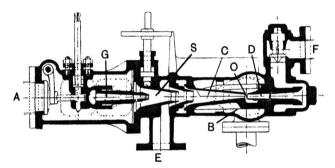

Fig. 23.   Exhaust steam injector.

is capable of thus developing a pressure of 120 lbs., and for the additional pressure required to force the water into the boiler a small jet of live steam is introduced through a supplementary nozzle.

Another form of injector using live steam from the boiler has warming cocks fitted, so as to enable the driver to blow surplus boiler steam into the water tank whenever the safety valves are lifting;

most drivers, however, prefer to keep the water in
their tenders cold, especially if there is any doubt
as to the ability of the injector to deal with hot water.

No doubt hot feed will become more extensively
used when locomotive superintendents are thoroughly
convinced of the modern injector's capability to pass
hot water with the same certainty with which it
takes cold water.

The chief and most important means of increasing
the efficiency of the steam is by superheating which,
together with the methods of feed heating and
thermal storage, will claim our attention in the
next chapter.

CHAPTER IV

SUPERHEATING, THERMAL STORAGE, FEED
HEATING

THE effect of heat upon water is to convert it into
steam. That portion of the heat which produces the
necessary rise in temperature is called the *sensible*
heat. Thus, to raise the temperature of one pound
of water from 32° F. to 212° F. or through 180°
requires practically 180 British Thermal Units*.

* In the production of one B.T.U. it is usually stated that 772 ft.-lbs.
of mechanical energy disappear. Later investigations, however, give
774 and 778. The original figure is accurate enough for all ordinary
investigations.

Or $h = t° \text{ F}. - 32°,$

where $h = $ the sensible heat.

After having reached the boiling point the water gradually disappears until the whole of the 1 lb. of water has been converted into 1 lb. of steam, during which process the temperature remains constant at 212° F. The heat thus imparted to produce the change of state, without change of temperature, is called *latent* heat. The conversion of 1 lb. of water to 1 lb. of steam absorbs 967 B.T.U. Note however that this quantity is true only for steam formed at the pressure of one atmosphere.

The latent heat may be approximately obtained from the formula

$$L = 1114 - 0·7 \, t° \text{ F}.$$

where $L = $ the latent heat in thermal units of one pound of steam formed at a temperature $t°$ F.

Steam in contact with the water from which it is generated is known as *saturated* steam and is steam at its maximum density. After the water has completely disappeared, if heat be still applied, the temperature as before will rise, provided the pressure is maintained constant : it is then known as *super-heated steam*.

Saturated steam is that used generally in loco-motive work : more recently superheated steam has been used. To understand the properties of both a knowledge of the relation between pressure, temperature, and volume is essential. These relations have been obtained by Regnault from experimental data and the values met with in locomotive boiler working are given in round numbers in the following table.

Properties of Saturated Steam

Gauge Pressure of boiler (lbs. per sq. inch)	Temperature (° F.)	Total Heat (Thermal Units)	Latent Heat (Thermal Units)	Volume of 1 lb. (in cub. feet)
150·3	365·7	1193·5	855·1	2·72
160·3	370·5	1194·9	851·6	2·58
170·3	375·1	1196·3	848·2	2·45
180·3	379·5	1197·7	845·0	2·33
190·3	383·7	1199·0	841·9	2·22
200·3	387·7	1200·2	838·9	2·12
215·3	393·6	1202·0	835·8	1·98
225·3	397·3	1203·1	833·1	1·9

It may be stated generally that the pressure varies with the temperature, the rate of change of pressure increasing more rapidly as the temperature increases. A formula expressing the connection between the temperature and pressure of saturated steam given by Rankine is as follows:

$$\log p = 6\cdot1007 - \frac{B}{T} - \frac{C}{T^2},$$

in which
$$T = t + 461° \text{ F.}$$

(the formula for converting the Fahrenheit scale to the scale of absolute temperature),

$$\log B = 3\cdot4364,$$
$$\log C = 5\cdot5987.$$

For all ordinary purposes in connection with locomotive investigation, however, the tables suffice. The connection of pressure and volume is usually expressed by the formula (also by Rankine):

$$PV^{\frac{17}{16}} = 479,$$

where

P = pressure in lbs. per sq. inch,

V = volume in cubic feet per pound of steam.

The *total heat of steam* is the total of the *sensible* and *latent* heat required to raise the temperature of one pound of water from 32° F. and convert it into saturated steam at any given temperature. Thus, according to definition,

$$H = h + L,$$

where

H is the total heat,
h the sensible heat, and
L the latent heat.

But we have seen that L may be expressed

$$1114 - 0\cdot7\, t°, \quad \text{and} \quad h = t° \text{ F.} - 32°,$$

whence we get

$$H = (t° \text{ F.} - 32°) + (1114 - 0\cdot7\ t°)$$
$$= 1082 + 0\cdot305\ t° \text{ F.}$$

The factors H and L are given in the table, and h may be obtained by subtracting the figure in column 4 from that in column 3.

Now locomotive steam is generally very 'wet,' i.e. it contains suspended moisture, due to the violent ebullition and the small water surface available for the steam to escape from. The dryness fraction which is used to express this condition of the steam averages about 10 per cent. The presence of this moisture means that less heat is required than is necessary to produce the same weight of dry steam, but this is no advantage since wet steam is not only very undesirable in the cylinders, but represents coal burnt to no purpose. It is obvious then that dry steam would mean a considerable saving. This and more is obtained by using superheated steam.

We have seen that superheated steam results from a continued application of heat to the steam after all the water has been evaporated. What happens is that its temperature then becomes more than that due to the pressure, a state impossible with saturated steam which has only one temperature for a given pressure. If, as we shall see happens in the engine cylinders, heat is abstracted from saturated steam, its temperature is not lowered but some of it is condensed into water. On the other hand the addition of heat at constant pressure, that is to say, under conditions which permit the steam to expand as it is heated, causes a rise in temperature. Such steam is no longer saturated but superheated. In this state and in proportion to its temperature rise it behaves less like a vapour and more like a perfect gas, one result of which is that its volume per pound also increases at a rate roughly proportional to the increase of its absolute temperature. Its temperature may also be reduced without condensation. Superheated steam has a greater volume than the same weight of saturated steam, the increase in volume being roughly $12\frac{1}{2}$ per cent. for every $100°$ F. of superheat added. Its specific heat does not appear to be constant, but for practical purposes it may be taken as equal to $0{\cdot}48$ at constant pressure.

To ascertain the total heat required to form superheated steam, the total heat of saturated steam at the given pressure is first found according to the equation stated above, to which is added the heat required to superheat the steam given by

$$0{\cdot}48\,(t_s - t_1),$$

where

   t_s = the temperature due to superheating,
   t_1 = the temperature of the boiling-point due to the pressure.

It is evident therefore that with superheating additional heat is required, which, however, is carried as an increased number of units per pound of steam to the cylinder with a very considerable effect upon efficiency. In the first place the loss occasioned by wetness carried over from the boiler is removed, and that due to initial condensation and heat interchange between the steam and cylinder walls is reduced to an extent dependent upon the degree to which super-heating is carried. The last mentioned losses need explanation. They are due to the action of the piston in a cylinder. As it moves up the cylinder the pressure of the steam is reduced by expansion, consequently the temperature is reduced. This means that condensation takes place to form water. The condensed steam is partly re-evaporated by the next inrush of steam, but this robs it of its heat, and so reduces its efficiency of work. This heat exchange is continually going on at every stroke of the piston, and, in fact, the formation of a film of water on the metal surface of the cylinder constitutes the heaviest loss in the expansive working of steam.

As superheated steam cannot become condensed until the temperature has fallen back to its saturation or boiler temperature, it becomes more stable, and it is thus possible to use the steam in the cylinders in a dry state without any losses due to liquefaction. A higher theoretical efficiency is thus obtained from

the steam owing to its greater elasticity; also, as
one effect of superheating is to increase the volume
occupied by a given weight of steam without altering
its pressure, a less weight of steam is required per
stroke. Prof. Ripper states that $7 \cdot 5°$ of superheat
are sufficient to compensate for loss due to 1 per cent.
of initial condensation, and he has shewn that the
heat exchange between the steam and cylinder walls
is correspondingly reduced.

It will probably occur to the reader that this is
all very well, but only a given quantity of heat can
be generated in the fire-box, and if a portion of this
is used for superheating, so much the less is available
for producing steam; in fact it appears to be a
question of taking a penny out of one pocket and
putting it into another. With this in view how
exactly is the increased efficiency to be accounted
for? No one has explained this more clearly than
Prof. Ripper*. Suppose an engine using saturated
steam with 25 per cent. of the steam condensed up to
the point of cut-off. Then since 1 per cent. of wetness
requires $7 \cdot 5°$ F. of superheat, 25 per cent. of wetness
will require $7 \cdot 5 \times 25 = 187 \cdot 5°$ F. of superheat. But
the specific heat of superheated steam is $0 \cdot 48$, hence
there is required $187 \cdot 5 \times 0 \cdot 48 = 90$ thermal units per
pound of steam. As there is only 75 per cent. of the
steam engaged in doing useful work, approximately

* The Steam Engine in Theory and Practice.

1000 heat units per pound of steam would be supplied, and putting the heat efficiency at 10 per cent. 100 out of the 1000 units are converted into work. By supplying, as above, 90 thermal units as superheat, the whole of the steam present in the cylinder is 'dry' and the useful work done is increased approximately in the proportion of from 75 to 100 = a gain of 33 per cent. This gives 133·3 heat units converted into work out of a total of 1090, or an efficiency of 12·23 per cent. as against 10 per cent. without superheat. The portion of the heat used for superheating thus shews the high efficiency of

$$\frac{33\cdot3}{90} \times 100 = 37 \text{ per cent.}$$

So much for the theory of the subject. When applied in practice we should expect to see this increased efficiency represented by a saving in coal and water for a given power. Numerous tests carried out on actual locomotives both when stationary and on the road shew that economy in fuel and water and increased efficiency are so obtained and in proportion to the increasing degree of superheat. To cite one only of numerous elaborate locomotive tests, namely, that carried out by Prof. Goss at the Purdue University, it was found that the substitution of superheated for saturated steam for a given fixed power permits:—

The use of comparatively low steam pressures, a

generally accepted limit being 160 lbs.: a saving of from 15 to 20 per cent. in the amount of water used: a saving of from 10 to 15 per cent. in the amount of coal used while running, or of from 3 to 12 per cent. in the total fuel supplied : assuming the power developed to equal the maximum capacity of the locomotive in each case, the substitution of superheated for saturated steam will permit an increase of from 10 to 15 per cent. in the amount of power developed, accompanied by a reduction in total water consumption of not less than 5 per cent. and by no increase in the amount of fuel consumed.

In tests carried out in actual working Mr Hughes, of the Lancashire and Yorkshire Railway, shewed that a superheater engine, when put to the highest test, that is by running against a compound engine, gave results which represent an economy in total coal per train-mile of 12·6, and per ton-mile of 12·4 per cent. in favour of the superheater.

Again, to take the most recent results available, Mr Bierman, of the Dutch Railway Company, gives as the results of carefully made runs with express and ordinary trains a saving of 2·17 kgs. per train-kilometre (7·70 lbs. per train-mile) representing for the seven months which the locomotives were employed on trial, a saving of 377,455 kgs. (832,145 lbs.) of coal in running 173,972 locomotive-kilometres (108,103 locomotive-miles).

The impulse was first given to the now widely prevailing movement of superheating by its re-introduction in 1898 on German engines. [It is not generally known that as far back as 1845, a Gt Western engine was fitted out with a superheater and that in the early fifties MacConnell, on the London and North Western Railway, also employed superheaters on some of his locomotives. These experiments were apparently in advance of their time.] Belgium, France, Switzerland and America followed to the extent that the practice has, in combination with compounding, become standard in these countries. British engineers were slower to convince, but after careful and tentative trials, the practice is steadily advancing, and locomotives so fitted are found on most of our leading railways.

As an example of a superheating apparatus employed in Great Britain, that introduced last year by Mr C. J. Bowen Cooke on the London and North Western Railway is illustrated in Fig. 24. The lower rows of tubes A, which carry the furnace gases from the fire-box to the smoke-box, are of the usual type and diameter. The upper rows B are much larger, and in these larger tubes, twenty-four in number, the steam superheater tubes are arranged. When the steam regulator valve D is opened the steam passes from the boiler along the main steam pipe E to a steam collector F fixed on the front of the boiler.

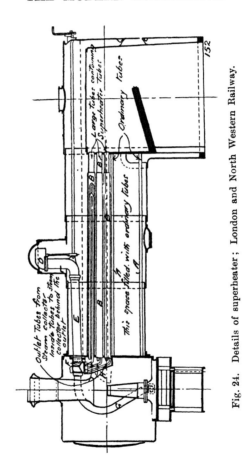

Fig. 24. Details of superheater; London and North Western Railway.

The steam collector is divided into compartments, a
saturated steam chamber receiving the steam direct
from the boiler at a temperature of about 377° F.,
and the superheated steam chamber receiving the
steam from the superheater tubes at a temperature
of about 650° F., passing from which it passes to the
cylinders.   One end of each superheater tube is
connected to the saturated steam chamber, whence it
runs along the large tube B, nearly up to the fire-box
and back to the superheated steam chamber, to
which the other end is connected.   The steam on its
way from the boiler to the cylinders thus passes
through the regulator valve to the steam collector,
through the superheater tubes and back to the
steam collector, and thence by the steam pipe to the
cylinders.   It becomes superheated to a maximum of
about 650° F.

The temperature of the superheated steam is
measured by a pyrometer connected to the superheater chamber of the steam collector, and is indicated
by a gauge fixed in the engine cab under the obser-
vation of the engine driver.   In order to regulate the
amount of superheat a movable plate H is fixed on
the smoke-box tube plate, or front of the boiler
barrel, by means of which the temperature of the
heated gases passing through the large fire tubes
may be controlled, and the temperature of the steam
passing though the steam tubes within them regulated.

Feed-Water Heating. The method almost invariably resorted to in stationary engine practice of
utilizing the residuum of heat in the exhaust steam
or the flue gases after leaving the flues, for heating
the water fed into the boiler, has not yet found
extensive imitation in locomotive design.

This is probably due to the fact that until recently
it was not possible to adapt the injectors to the work
of feeding hot water, and that English locomotive
practice had discarded feed pumps in favour of
injectors. Feed pumps, successfully to replace injectors, must be independent steam-driven units.
These, however, take up room, and perhaps necessitate
more attention from the driver. Nevertheless, Mr
Drummond on the London and South Western Railway has tackled the problem vigorously and fitted a
number of his engines with feed-water heaters. The
apparatus is illustrated in Fig. 25. The exhaust steam
from the pumps which deliver the hot feed-water to
the boiler is sent to the tender with that portion of
the main exhaust utilized for the purpose. The water
is pumped into the boiler at a temperature of about
180° F. Mr Drummond states that a saving in fuel
equal to 6 lbs. per mile is effected. The tank from
which the feed is immediately drawn, and through
which the exhaust steam heating pipes are led, is
supplementary to the tender tank. The condensed
water escapes through a series of holes in the rear

casting which receives the ends of the pipes, whilst
the uncondensed steam passes into the atmosphere
through an escape pipe at the rear of the tender.

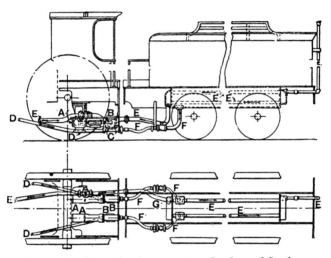

Fig. 25. Feed-water heating apparatus ; London and South
Western Railway.

A   Steam cylinder (5½ ins. diam.
    9 ins. stroke).
B   Pump (4½ ins. diam. 9 ins.
    stroke).
C   Valve box.

D   Delivery from pump.
E   Exhaust steam from cylinder.
F   Pump suction.
G   Exhaust steam from pump.

Thermal Storage. Mr Druitt Halpin's system
of thermal storage as applied to steam boilers for
stationary engines has, in certain circumstances,

shewn itself to possess distinct advantages over
ordinary methods of boiler feeding. A test conducted
by Prof. Unwin with Cornish boilers shewed a coal
saving of 19·7 per cent. In order to ascertain the
increased efficiency, if any, due to the application of
the system to locomotives, Mr Ivatt, when locomotive
superintendent of the Gt Northern Railway, fitted
a 2–4–0 type passenger engine with the Halpin
apparatus. The arrangement is very simple, and
consists of a cylindrical storage tank placed above
and connected to the boiler by means of a pipe. All
the feed-water, which is maintained at or about the
same temperature as the water in the boiler, is passed
through the cylinder, the water being heated by
steam generated during the intermittent periods when
the engine is standing or the safety valves are blowing.
The water thus heated is fed to the boiler as required
when the engine is running, this being regulated by
a valve in the driver's cab. Six tank engines on the
Lancashire and Yorkshire Railway were some time
ago equipped with this apparatus. Where stopping
places are frequent and on rising gradients Mr Hughes
states that there is distinct economy. Certain tests
were carried out between Salford and Accrington,
resulting in an actual saving of 1 ton of water, and
under similar conditions elsewhere the saving was 12
per cent. When, however, these engines have to take
their turn on other sections of the line which are not

so favourable, the all-round economy is brought down to 4 per cent.

CHAPTER V

RESISTANCE, TRACTIVE EFFORT, ADHESION

HAVING seen how the steam is generated the question arises, what work is to be done by it? The engine has not only to propel itself but to overcome the resistance offered by the train. The combined resistance is made up of several components. (1) Resistance dependent on the speed. (2) The resistance caused by flange action and weather. (3) Resistance due to gradient. (4) Rolling and axle friction and side play. Resistance dependent on the speed is due to the friction of the mechanism of the engine and the air resistance due to engine frontage. The determination of this is still a subject of investigation, and various formulae are proposed from time to time, the results obtained from which, however, do not appear to agree very closely amongst themselves. Mr Daniel Gooch, of the Great Western Railway, conducted a number of experiments during the gauge controversy, from which D. H. Clark obtained the much used formula

$$R = 8 + \frac{V^2}{171} \quad \ldots\ldots\ldots\ldots\ldots(1),$$

where

V = the velocity in miles per hour

R = train resistance in pounds per ton

for engine and vehicles combined, which is based on the assumption that the rolling stock and rails are in good condition, and assuming an absence of side wind and wet. Lubrication at that period (1855) was effected with grease, which has since been replaced by oil, thereby reducing axle friction from about 6 lbs. per ton to between 3 and 4 lbs. at slow speed, the resistance rising as the speed increases. To meet this the formula (1) was modified to

$$R = 6 + 0\cdot009\,V^2 \quad \ldots\ldots\ldots\ldots(2).$$

It may be noticed too that so far as the interaction between wheel and rail is concerned, rolling friction has been reduced by the adoption of steel for tyres and rails.

A good working formula proposed by Pettigrew is

$$R = 9 + 0\cdot007\,V^2\ldots\ldots\ldots\ldots\ldots(3).$$

Later still M. Barbier of the Chemin de Fer du Nord presented formulas which in construction have been followed by nearly all other investigations.

The following table gives the most important results in tabulated form.

Formulas for Train Resistance

R = Tractive resistance in lbs. per ton (2240 lbs.).
V = Speed, miles per hour.
L = Length of train in feet.

No.	Authority	Formula	Remarks
1	Clark	$8 + \dfrac{V^2}{171}$	Whole train
2	Sinclair	$2 + 0 \cdot 24 V$	—
3	Pettigrew	$9 + 0 \cdot 007 V^2$	—
4	Deeley	$3 + \dfrac{V^2}{290}$	—
5	Barbier	$3 \cdot 58 + 1 \cdot 658 V \times \left(\dfrac{1 \cdot 609 V + 50}{1000} \right)$	4-wheeled Vehicles
6	,,	$3 \cdot 58 + 1 \cdot 644 V \times \left(\dfrac{1 \cdot 609 V + 10}{1000} \right)$	Bogie Vehicles
7	,,	$8 \cdot 51 + 3 \cdot 24 V \times \left(\dfrac{1 \cdot 609 V + 30}{1000} \right)$	Engine and Tender
8	Aspinall	$2 \cdot 5 + \dfrac{V^{\frac{5}{3}}}{50 \cdot 8 + 0 \cdot 0278 L}$	Bogie Coaches

Resistance due to Gradient. In addition to over-coming the friction of the mechanism, the engine must be able to haul its load up inclines. The effect of gravity against ascending an incline can be ex-pressed by,

$R = WX \sin \theta$, where

R = the resistance in lbs. per ton hauled.

$$W = \text{the load and } \sin \theta = \frac{\text{vertical rise}}{\text{length of incline}}.$$

Thus if $W = 1$ ton and $\sin \theta = \frac{1}{300}$ then

$$\frac{1 \times 2240}{300} = 7\cdot4 \text{ lbs. per ton due to gravity.}$$

In comparison with axle friction this represents a factor which does not admit of reduction.

It is seldom possible to ascertain the actual weight of a train, but if the number of axles be counted and 5 tons allowed for each, a very fair estimate of the weight of a passenger train can be made.

Resistance due to Curves. When a train runs through a curve, especially if it be a reverse, or S-curve, a large amount of resistance is set up by the grinding action of the wheel flanges against the rails, the collars of the axle journals being forced against the bearings, thus developing end friction. Curve resistance depends upon the radius of the curve and the length of the rigid wheel base of the vehicles. It is a rather uncertain quantity involving the state of the rails, whether dry or greasy, and the strength and action of the wind. A formula due to Morrison is

$$R = \frac{WF(D+L)}{2r},$$

where R = resistance; W = weight of vehicle; F = coefficient of friction between wheel and rail varying according to weather from 0·1 to 0·27 ; D = distance of rail between treads ; and L = length of rigid wheel base.

Much has been done in recent years to reduce curve friction by the provision of better arrangements for end wear, lubrication and short-based bogies. Increased resistance and wear are occasioned by large flange play. The wind has a great effect in increasing train resistance. A head wind virtually increases the velocity with which the train travels against the air. This resistance reaches a maximum when the wind is blowing at right angles to the train and produces the side effect similar to that on a curve. Carus Wilson states that the resistance of the air with a train of bogie coaches running at 60 miles per hour, amounts to about one half of the total tractive effort required to haul the train. It is claimed by some that a large reduction can be made by the adoption of wedge-shape 'wind cutters,' familiar on Bavarian locomotives, to the extent of 10 per cent. of the total tractive effort with a passenger train. Against this, however, must be set the fact that when the engine is running round a curve, or is exposed to a side wind, the air pressure, so far from being reduced, is intensified.

Resistance to the progressive movement of a train may be determined, when uniform speed has been attained, by calculating the total force exerted by the aid of indicator diagrams ; then deducting the drawbar pull, as denoted by a dynamometer, we have for difference the total resistance of the loco-motive alone. A second method of determination is to shut off steam at any given point and to calculate the operative force from the speed variation. In applying this method the engine is usually allowed to come to a standstill on a downhill gradient, and the resistance to motion is equal to the retarding force plus the acceleration due to the gradient.

The question of resistance to locomotives running at high speeds is of a complex character, for in addition to the commonly recognized forces causing resistance there are others of more obscure character which, being apparently developed within the machine, give rise to what is called 'internal resistance.'

It is known that the size of the wheels and the arrangement of mechanical features have a very important effect on the running of an engine. This point has been well illustrated by Mr Ivatt, when he was chief at Doncaster, by means of diagrams taken from Great Northern engines shewing the relation between horse-power and drawbar pull.

Indicator diagrams shew the power developed in the cylinders, but not the proportions of the total power exerted in the form of drawbar pull, because— and particularly at high speeds—much of the cylin-der power is absorbed in overcoming the internal resistance of the engine itself. With increase of speed, internal resistance increases and drawbar pull diminishes, until a point is reached at which the engine is only able to move itself and exerts no pull at all on the drawbar. This will be more fully realised by an examination of the following figures given by Mr Ivatt.

Comparison of Drawbar Pull for Two Locomotives at Different Speeds

Speed in Miles per hour	Drawbar Pull in Tons	
	Eight-coupled Goods engine	Single-wheeled Express engine
10	7·6	(3·8) *
20	4·6	3·0
30	2·0	2·5
40	0·9	2·1
50	(0·1) *	1·8
60	—	1·3
70	—	0·8
80	—	0·4

* Computed.

While simply illustrating the behaviour of two extreme types of engine, the table helps to shew the advantage to be derived from what is termed a 'free running' engine.

The amount of power absorbed by a locomotive is something astonishing to the uninitiated. According to Mr Sisterson the power absorbed in running an engine weighing from 80 to 90 tons, together with its tender, amounted to between 800 and 900 I.H.P., when the speed of about 70 miles on the level was attained. This represents a resistance of very nearly 60 lbs. per ton of engine and tender. Taking another example, based on the running of the *Precursor*, a 4–4–0 type of engine designed by Mr Whale in 1905 for the London and North Western Railway, it was found that during a run between Crewe and Rugby at 61 miles an hour, the drawbar pull was 2 tons, equivalent to about 730 horse-power while the engine was developing 1174 horse-power. Here we have $1174 - 730 = 444$ I.H.P. representing resistance of the engine alone. It would be interesting to know exactly what becomes of such very considerable amounts of power, but no one is prepared with a precise explanation.

Inertia. When a train is started from rest an accelerating force is required to put the mass of the train in motion in addition to the force required to overcome frictional resistance. This, however, is independent of the uniform rate of motion considered above, and applies more particularly to suburban tank engines.

Adhesion. Closely connected with the load drawn

is the adhesion between the driving wheels and the rail, that is to say, the friction between them available to resist slipping. If the adhesion is not at least equal to the resistance the wheels will rotate and slip on the rail without advancing. The adhesion is equal to the weight on the driving wheels multiplied by a coefficient which depends upon the condition of the surface of the rail. This may vary between $\frac{1}{3}$ in dry weather, to $\frac{1}{10}$ in wet when the rails are greasy. It is sufficiently accurate to take the value $\mu = \frac{1}{4}$. The weight on the driving wheels depends on the wheel arrangement adopted. With the single, 2–2–2 type, only the weight of a single pair of wheels is utilized, and as the strength of the rail imposes a limit of 20 tons, this represents the limit of adhesion of the single engine. In the 4–4–0 and 4–6–0 types two-thirds or more of the total weight of the engine is available for adhesion, and in the case of goods engines of 0–6–0 and 0–8–0 types the whole of the weight is so utilized. Thus, the resistance to be overcome is a determining factor of the wheel base.

Tractive Effort. To overcome the total resistance of the train the tractive effort produced by the action of steam on the piston by which propulsion is determined must at least equal it. Let the area of the piston be $\dfrac{\pi d^2}{4}$, the stroke $= l$ and the mean effective pressure $= p$; then the work done by the cylinder will be $p\,\dfrac{\pi d^2 l}{4}$, and for one revolution of the wheel, in a two-cylinder engine, $p\pi d^2 l$. Let $E =$ the mean effort necessary to propel the engine and train ; and the distance travelled during one revolution of the wheel πD, D being the diameter of the driving wheel, the work done is then $\pi D E$. Equating these two values we get

$$\pi D E = p\pi d^2 l,$$

whence $E = \dfrac{d^2 l p}{D}\,.$

This value E represents the mean tractive effort of the locomotive ; the mean pressure p is only a fraction of the boiler pressure and must be evaluated.

Thus for an engine with cylinders 18 ins. in diameter by 24 in. stroke 6 ft. driving wheels, and taking the mean effective pressure at 80 per cent. of the 200 lbs. the boiler pressure

$$\text{Tractive force} = \frac{18 \times 18 \times 24 \times {\cdot}8 \times 200}{72}$$
$$= 25,920 \text{ lbs.}$$

In an engine working compound (see chapter on Compounding) the tractive effort is thus determined. Let p and p_1 be the mean effective pressures in the high- and low-pressure cylinders respectively, d and d_1 the respective diameters, l the stroke common to both. In a two-cylinder compound engine the work per revolution is

$$\frac{\pi d^2 l}{2} p + \frac{\pi d_1^2 l}{2} p_1,$$

and the tractive effort

$$E = \frac{p d^2 l + p_1 d_1^2 l}{2D}.$$

In a four-cylinder engine, the factors d^2 and d_1^2 must be replaced by $2d^2$ and $2d_1^2$ since there are two high- and two low-pressure cylinders.

Therefore we obtain

$$E' = \frac{p d^2 l + p_1 d_1^2 l}{D} \quad \dots\dots\dots\dots(1).$$

The above formula involves the determination of the mean effective pressure.

Von Borries has given the following rule for a two-cylinder compound (the result must be multiplied by 2 for a four-cylinder engine):

$$d^2 = \frac{4T \times D}{p \times S},$$

where $d =$ Diameter of the low-pressure cylinder,
$T =$ Tractive effort,
$D =$ Diameter of driving wheel,
$p =$ Boiler pressure,
$S =$ Stroke of piston,

whence $$T = \frac{d^2 \times p \times s}{4 \times D} \quad \dots\dots\dots\dots(2).$$

The formula used by Baldwin for estimating the tractive power of four-cylinder compounds, is as follows:

$$T = \frac{C^2 \times S \times \frac{2}{3}P}{D} + \frac{c^2 \times S \times \frac{1}{4}P}{D} \quad\ldots\ldots\ldots(3),$$

in which C = Diameter of H.P. cylinder in ins.,
c = Diameter of L.P. cylinder in ins.,
S = Stroke in ins.,
P = Boiler pressure in lbs.,
T = Tractive power,
D = Diameter of driving wheel in ins.

Another formula is

$$T = \frac{1\cdot6 pr^2 s}{D(2+1)} \quad\ldots\ldots\ldots\ldots\ldots(4),$$

whence r is the ratio of the cylinder volumes, the other equivalents being as in (2).

Mean Effective Pressure.   In the locomotive as indeed in all steam engines, the steam is used expansively.   Steam is admitted during the period the piston is performing a portion of its stroke, and the valve then closes, cutting off the steam.   The steam in the cylinder then expands, expansion continuing and the pressure diminishing until the piston has nearly completed its stroke when the exhaust takes place, and the pressure falls very nearly to that of the atmosphere.

During admission the pressure is practically uniform, and from the point of cut-off until the exhaust commences expansion follows very closely Boyle's law : pv = a constant.

In the locomotive the point of cut-off is arranged to take place from 75 per cent. of the stroke down to 20 per cent., according to the nature of the work required. The average or mean effective pressure on the piston can be determined either from an indicator diagram or by calculation.

Readers are referred to a text-book on the steam engine for an explanation of the indicator and its method of use. It will suffice to state here that, by this apparatus, a figure or diagram is traced on a piece of paper representing the pressure of the steam in the cylinder ; the upper line shews the pressure urging the piston forward and the lower line the pressure retarding its movement on the return stroke. The mean effective pressure may be obtained by calculation from the equation

$$P_m = P_1 \left(\frac{1 + \log_e r}{r} \right) - p_2,$$

where

P_m = Mean effective pressure.

P_1 = The boiler pressure, plus that due to the atmosphere = 15 lbs.

$p_2 =$ The back pressure plus that due to the atmosphere = say 19 lbs.

$r =$ The ratio of expansion calculated by dividing the volume of steam in the cylinder at the end of the stroke by the volume of steam in the cylinder at the point of cut-off, i.e. by dividing the length of stroke by the cut-off. It may be put at 1·33 for 75 per cent., and 5 for 20 per cent., of cut-off respectively.

$\log_e r =$ the hyperbolic logarithm of r, the ratio of expansion. For $r = 1·33$ and $r = 5$ the hyperbolic logarithms are 0·285 and 1·609 respectively.

For example. Let the pressure be 175 lbs. = 190 lbs. absolute. Then for 75 per cent. cut-off

$$P_m = 190 \frac{(1 + 0·285)}{1·33} - 19 = 164·5 \text{ lbs.}$$

For 20 per cent. cut-off

$$P_m = 190 \frac{(1 + 1·609)}{5} - 19 = 80·1 \text{ lbs.}$$

The following table gives a few hyperbolic logarithms required in locomotive practice.

Hyperbolic Logarithms

Ratio of Expansion	Hyperbolic Logarithms
1·35	0·3001
2·0	0·6931
2·5	0·9168
3·0	1·0986
3·5	1·2528
4·0	1·3863
4·5	1·5041
5·0	1·6094
6·0	1·7918
7·0	1·9459

CHAPTER VI

UTILIZATION OF THE STEAM

THE conversion of the energy of the steam into the work necessary to overcome resistance and thus propel the engine itself and its load is accomplished in the cylinders. The cylinder is a cast-iron casting the interior of which is truly bored out to cylindrical shape, to afford a smooth surface for the recipro-cating motion of the piston. To render the piston steam-tight, grooves are turned in its edge into which are sprung elastic rings made of steel which tend to press outwards against the cylinder walls. A piston

rod is attached to the piston by means of a nut fitting
a screw on the rod. The end of the rod is tapered
off to pass through a tapered hole in the piston which
thus prevents it becoming slack on the rod. The rod
passes through a hole in the front cylinder cover, the
joint being made steam-tight by means of a stuffing
box containing metallic packing. The reciprocating
motion of the piston and its rod is converted in a
rotary motion at the crank axle, the necessary connection being made by the connecting rod. As there
are usually two cylinders, there are thus two cranks.
These are set at an angle of 90° to each other, so that
when one piston is at the end of its stroke, or on the
dead centre, the other is in its position of maximum
effort. Steam is admitted alternately on opposite
sides of the piston through two steam ports, one at
each end of the cylinder, leading from the steam
chest. A third port, called the exhaust port, allows
the steam to escape to the blast-pipe. These ports
open into the steam chest in which the slide valve
reciprocates and so distributes the supply of steam to
the ports and thus to the piston. The valves are
driven by a valve gear or motion driven by eccentrics
on the main shaft, or by other means which we shall
examine later. The cylinders are arranged at the
front of the engine generally under the smoke-box
and either inside or outside the frames. The prevailing British practice is to place them between the

frames, which method imparts a rigidity to the whole
structure since the cylinder casting itself serves as
a frame stay. Further, the effort set up by the steam
and moving parts acting at a minimum distance
from the longitudinal axis of the engine, a greater
steadiness in running is obtained. Foreign practice
generally, however, favours the outside cylinder
arrangement in that it permits the use of larger
diameter cylinders, ready accessibility of the parts
and the elimination of the cranked axle.

In compound engines three and four cylinders
are employed, the high-pressure cylinders being
arranged outside and the low-pressure inside the
frames. In the latest engines the cylinders are
stepped, that is, one pair is set in advance of the
other.

The steam chests occupy a position corresponding
to the type of valve gear employed. With interior
cylinders they are placed either between, above, or
below the cylinders; with outside cylinders the gear
is also generally outside and the steam chests placed
on top of the cylinders, sometimes horizontally and
sometimes inclined towards the exterior. The outside cylinder engines of this country have the valve
gear and steam chests disposed inside the frames:
in American engines the steam chest is placed outside, above the cylinder, communication between
the valve rod and valve gear being made through a

rocking shaft. Cylinders are always made from a
hard close-grained cast iron and when of the inside
type, are generally cast in pairs. Quite recently the
method of casting them *en bloc* has been adopted
thus doing away with a joint and increasing the
rigidity.

Pistons are usually of cast steel with cast iron or
cast steel piston rings, which, when in position, are
about $\frac{1}{32}$ in. open.

Slide Valves. Valves are either of the flat, or
'D' type, or cylindrical in shape when they are
known as piston valves. Various modifications of
the old flat valve have been introduced in recent
years with the object of reducing the excessive
friction between the valve and valve face. With
these the steam exerted its full pressure on the
whole area of the valve back, with the result that
a large percentage of the power developed in the
cylinder was required to move it. With flat valves,
what is called 'balancing' is now largely resorted to,
one of the latest designs of a valve so modified being
shewn in Fig. 26.

The main valve consists of three principal parts;
the valve proper AE, the balance plate, S, and the
pressure plate above it.

The valve has two faces, one operating on the
valve seat on the cylinder, and the other against the
face of the balance plate. Both faces are the same,

and that of the balance plate against which the valve
operates is a duplicate of the cylinder valve seat.
The walls of the valve are provided with ports AE,
which pass from face to face of the valve. On the
opening of a steam port the pressure has free access
to both sides of the valve by reason of the passages
AE through the valve to the port F in the face of
the balance plate, which corresponds with the cylinder
port. Consequently the pressure in the port has no

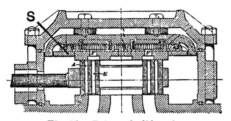

Fig. 26.   Balanced slide-valve.

effect upon the valve as it acts on both sides of the
valve face in equal area and pressure.

The piston valve was also introduced to reduce the
frictional resistance to the valve movement. Briefly
it consists (Figs. 10 and 27) of a hollow cylinder turned
at the ends to fit a bushing in which the steam ports
are cut. It is reduced in section in its central portion.
The ends are fitted with L-shaped packing rings,
similar in construction to the piston rings of a
cylinder, and uncover the steam ports to steam and

exhaust at the proper time. The motion of the valve
and steam distribution are the same as in the D
valve, but as the pressure does not act in forcing it
up against the walls of its bushing or seat, it is easily
driven. There is just enough area of ring to make it
steam tight without unnecessary friction. The body
is usually made hollow of cast iron : in the latest
practice seamless steel tubing is employed with light
cast steel ends riveted on.

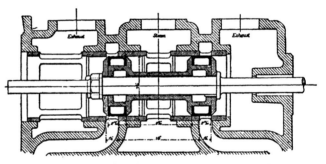

Fig. 27.   Piston valve ; Lancashire and Yorkshire Railway.

The greatest disadvantage under which the piston
valve labours is its inability to relieve excess pressure
in the cylinder port by lifting, after the manner of
the slide valve. This renders the employment of a
cylinder relief valve imperative. To eliminate the
disadvantages attaching to the use of such valves,
Mr Hughes on the Lancashire and Yorkshire Railway

has adopted auxiliary valves on the piston valve itself. These are held on their seats as long as the pressure in the cylinder does not rise beyond the working pressure. Should, however, an excess of pressure arise in the cylinder, it displaces these valves and the steam is discharged.

Valves of the lifting or 'poppet' type have now a considerable vogue on Continental locomotives. One

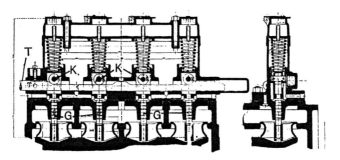

Fig 28. Poppet valve gear.

such, the Lentz, the details of which are shewn in Fig 28, was designed more particularly to meet the conditions set up by the use of highly superheated steam and to get rid of a complex mechanism. Each valve is screwed on to a steel spindle which moves up and down in a cast-iron guide and is rendered steam-tight by means of turned grooves, G, thus rendering the use of stuffing-boxes unnecessary. The spindles

7—2

end in broad cylinder heads in which rollers, *K*, are
arranged in such a manner as to turn easily. The
spindle heads slide up and down in their guides with
the valves and isolate the upper part of the case from
the lower chamber. The upper ends of the spindle
heads carry springs for loading the valves to ensure
the positive closing of the cam-rod by the rollers.
The cam-rod, *T*, is coupled to the valve gear instead
of the slide valve rod, and, in moving to and fro,
opens and closes the four different inlet and outlet
valves by means of the cams on which the rollers,
turning freely in the spindle head, move.

With the object of reducing the waste of heat
that occurs in ordinary locomotives with the reversal
of direction in the flow of steam, the Stumpf valve
gear has been introduced. With this apparatus
the steam flows in a continuous direction through
the cylinder, the inlet valves being arranged at the
ends, and does not cool the walls except at the centre
of its length, which is the part at which the exhaust
steam escapes.

The admission valves are of the double-seated
type with springs. Steam is admitted almost directly
against one of the faces of the piston, and, during
the latter portion of the piston travel, it escapes
through ports at the centre into the exhaust pipe.
The opening of the valve is insured by a mechanism
similar to that used with the Lentz valve.

An interesting type of gear in use on the Chicago and North Western Railway is of the rotary type and represents an adoption of the Corliss principle to suit the requirements of locomotive practice. Two valves are fitted to each cylinder, operating alternately as inlet and outlet, and driven by Corliss wrist motion. The valve gear makes use of the Stephenson link, eccentrics and rocker arm as far as the end of the valve stem. This is connected to a wrist plate, which has hinged attachments to a crank arm on the rotary valve spindle. A horizontal shaft extends across the back of the cylinder saddle, and this shaft is fitted with two cranks which connect with the bearings of the wrist plates. From the centre of the shaft a long connecting rod extends back to a short crank, so that when the link is raised or lowered from the central position, the wrist plate is raised to regulate the lead. The valve body is journaled in the heads of the steam chests, and its weight is supported entirely clear of the valve seat. The valve friction is thus materially reduced. The valve spindles, in their passage through the end of the steam chests, have a shoulder which forms a steam-tight bearing and requires no packing. This type of valve besides being efficient is stated to cause very little wear on the machinery.

Valve Gear. Of scarcely less importance than the boiler in determining the efficiency of the locomotive as a whole is the valve motion, upon which depends the proper distribution of the steam. The system most commonly employed is the gear known as the Stephenson, although it was in reality invented by William Howe. Others extensively used and closely resembling it are the Gooch and Allan. The most modern systems are, however, the Walschaerts and Joy motions. The Stephenson link gear is also known as the shifting link motion; the Gooch is directly opposite in its action, in that the link is stationary, and the link block attached to the valve rod is moved up and down. The Allan is a combination of the Stephenson and Gooch in that both the link itself and the valve rod are shifted. All of these motions are operated with two eccentrics, one for the forward and the other for the backward motion.

In the Stephenson and Allan motion when the eccentric rods are open, the lead* is increased as the

* For a detailed explanation of the terms 'lead,' 'lap,' the reader is referred to a text-book on the steam engine. It may be stated here, however, that *lead* is the amount of opening of the steam port at the beginning of the stroke of the piston. *Lap* is the cover of the steam ports by the outside, or steam, edges of the valve when the latter is at its mid travel. It represents the distance which the valve has to move from its middle position to open either steam port. The function of *lap* is to give a varying point of cut-off and so take advantage of the expansive quality of the steam. The simplest form of valve gear is the eccentric and its rod. The eccentric is in reality a crank whose pin is so enlarged as to envelope the shaft. Its eccentricity or 'throw' is the distance separating the centres of

link is pulled up and the point of cut-off made earlier.
If, however, the rods are crossed the 'notching up'
reduces the lead, though this reduction is much less
than the increase in the former case. Again with
either open or crossed rods, the corresponding increase or reduction of lead is much less with the
Allan than with the Stephenson valve motion. With
the Gooch, Walschaerts, and Joy motions the lead is
constant for all points of cut-off.

the eccentric and the shaft. As the crank shaft rotates the valve is
driven by the eccentric to and fro for a distance equal to twice the
throw. When the piston is at the end of its stroke, the valve will be
at half stroke just opening to steam and the eccentric is placed at 90°
to the crank (see Fig. 29). The eccentric leads the crank, otherwise
it would be closing the steam port when it should be opening it to
steam. Taking now a valve in its mid position with an outside lap

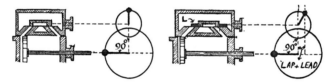

Fig. 29.   Position of a valve without lap or lead and of a valve
with lap and lead at the beginning of the stroke.

L (the portion shewn in black in Fig. 29). To uncover the steam
port it must be moved over a distance equal to L, and the crank,
being on the dead centre, the eccentric must lead the crank by 90° + lap
or distance L. But we have seen that when a valve has *lead*, the steam
port is already open at the beginning of the stroke. Thus the eccentric
must lead the crank by an amount equal to 90° + lap and lead.

104   THE MODERN LOCOMOTIVE   [CH.

The Walschaerts gear is driven by a combination of an eccentric or short stroke return crank from the main crank-pin and a connection to the crosshead. The Joy gear is driven from a connection to the connecting rod. The former has been extensively applied on the Continent of Europe and the latter is in use on the London and North Western, and Lancashire and Yorkshire railways.

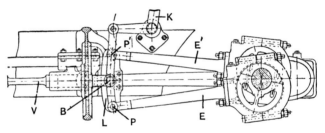

Fig. 30.   Diagram of Stephenson link motion.

As the Stephenson motion is the one mostly used in this country it will be first considered. It is illustrated in Fig. 30. *E, E* are the two eccentrics connected to a slotted link *L* at the points *P* and *P'*. The link is curved, the radius of curvature being that of the eccentric rod. It is capable of being raised or lowered by the lever *K* and accommodates in its slotted portion a block *B*, which slides in the slot. It is directly connected to the valve by the rod *V*. In the position shewn on the diagram the block, and therefore the valve, is not influenced by the motion of either eccentric and consequently the valve theoretically should not move. Actually however it moves very slightly. As the block occupies the top or bottom position of the link it is brought under the influence of either eccentric E' or E respectively and the valve travels its full distance. At intermediate positions the action of one eccentric is more or less neutralized by that of the other, consequently the travel of the valve is reduced. The extreme positions determine whether the engine will run in a backward or forward direction, the intermediate positions affect the distribution of the steam and the rate of expansion by altering the position of the cut-off. The *lead* and point of release of the exhaust steam are also thus capable of being varied.

The motion of a valve driven by link motion is thus exceedingly complex; it can, however, be approximately determined by geometrical methods, the explanation of which would be quite outside the scope of this book. The lever K is connected by a system of levers to a quick threaded screw, called the reversing screw, on the foot plate by which the reversal or cut-off is obtained.

Owing to the considerable manual effort required to operate the reversing gear of a modern locomotive, power reversing gear has been applied in many cases. One system, first introduced by Mr James

Stirling on the Glasgow and South Western Railway, employs a steam cylinder controlled and locked by a hydraulic cylinder or *cataract* gear. An example of that in use on the Great Eastern Railway worked by compressed air is illustrated in Fig. 31. The locomotive can be reversed in the usual way by means of the hand-wheel *A*, or by compressed air,

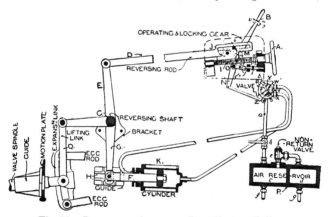

Fig. 31.   Power reversing gear; Great Eastern Railway.

the gear for this purpose being operated by the handle *B*. The reversing shaft *C* is connected to the reversing rod *D* by the arm *E* and to the piston rod *F* by the arm *G*. To operate the gear by power, the handle *B* is moved one way or the other, according to whether the engine is required to be put in

forward or backward running. This causes the valve
L to rotate, thereby opening one end of the reversing
cylinder K to pressure and the other end to exhaust.
In the running position both ends of the cylinder K
are in communication with the main air reservoir P
through the valve L, the piston rod F being made of
such a diameter that the reduced piston area, exposed to pressure on the piston rod side of the
piston, balances the weight of the motion hanging on
to the lifting links Q. Air is supplied from the
Westinghouse brake pump.

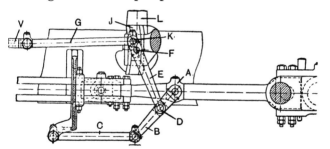

Fig. 32.   Joy's valve gear.

The Joy valve gear is an example of what is
known as the radial valve gear and is almost exclusively employed on the engines of the London and
North Western and Lancashire and Yorkshire railways. Some important advantages are obtained
with this gear, the chief of which is that the lead is

constant for backward and forward strokes and
remains so for all degrees of cut-off up to mid-gear.
The gear is illustrated in Fig. 32. The use of eccentrics is dispensed with. At a point *A* in the connecting rod is pivoted a link *B*, connected at its
lower end to the radius rod *C* which restricts its
motion to a vertical plane. The point *A* on the
connecting rod describes an ellipse when working.
At the point *D* a lever *E* is pivoted, which is centred
at *F* and extended to the point *K* where connection is made with the lever *G*. The path of the
point *D* is an irregular oval, and that of *K* a true
vertical ellipse. The valve rod *V* is pivoted to *G*.
A rising and falling movement is communicated to
F by the motion of the connecting rod, its movement
being guided by a die sliding in a slot *J* which has a
radius of curvature equal to the length of *G*. *G* rises
and falls with *F* and thus communicates a horizontal
movement to the valve spindle *V*.

The block in which the slot is cut is capable of
pivoting about a centre *F*, its inclination to one side
or the other being effected through the lever *L*
which is operated from the reversing screw at the
foot-plate.

The degree of port opening and consequently the
rate of expansion is regulated by the inclination of
the slot from the vertical. When it is exactly central
as shewn, the valve is in mid-gear; when thrown over

to its extreme backward or forward positions, backward or forward running of the engine is obtained.

The Walschaert valve gear, which is preferred on the Continent, has also a constant lead for all points of cut-off and produces a more uniform steam distribution than the Stephenson gear.

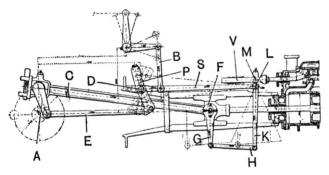

Fig. 33.   The Walschaert valve gear.

Referring to Fig. 33 it will be seen that the movement of the valve is derived from two sources, the crosshead F and an eccentric crank A, whose centre is situated 90° from the centre line on the main crank, when the centre lines of the cylinders and gear motion coincide and pass through the centre of the axle.   From the eccentric crank A an eccentric rod E runs and makes connection with the link D which is pivoted in the centre.

A groove in the link contains a die P, which is free to slide up and down therein: this block is attached to a radius rod S, the length of which between the points D and L is equal to the radius of the link itself. If the radius bar were pivoted directly to the valve spindle V, when the crank was on the centre, the valve would be in its mid position for either backward or forward running and there would be no lap or lead. Lap and lead are obtained by a rigid arm G, dropped from the crosshead centre F. Pivoted to G is a union arm H, making connection with the combination lever K. This is pivoted at the point L of the radius bar and prolonged to form a connection with the valve spindle at M. It will be seen that the inclination of the combination lever K will be the same at the end of the stroke regardless of the position of the radius rod S; and that, therefore, the horizontal displacement of the point M and the valve spindle will be the same on either side of a vertical line through L. This horizontal displacement is equal to twice the sum of the lap and the lead, hence the latter is constant for all points of cut-off. It should be pointed out that the eccentric and the crosshead tend to move the valve in opposite directions during the first half of each stroke and in the same direction during the last half; or, in other words, they work in opposite directions during the first and third quarters of a revolution of the crank

starting from either dead point, and together during
the second and fourth quarters. The motion derived
from the crosshead is constant and is not subjected
to reversal in the reversing of the motion of the
engine, which is done exclusively by a change in the
motion imparted by the eccentric crank, which also
controls the variation of the points of cut-off.

In order to accomplish this the motion of the
eccentric crank is transmitted through an oscillating
link pivoted at its centre and so slotted that a die
attached to the back end of the radius bar can be
moved through its whole length, and by placing this
above or below the centre a reversal of the engine
will be obtained. This motion, either direct or in-
direct, is taken up by the radius bar and carried out
to the combination lever, where it is combined with
that obtained from the crosshead and the resultant
imparted to the valve. The features which have
brought the Walschaert gear into such extensive use
abroad are its simplicity, lighter parts and accessibility, all of which are obtained without loss of
efficiency.

CHAPTER VII

FRAMES AND RUNNING GEAR

THE engine as a carriage does not shew so much divergence from the standard type of, say, twenty years ago as the other elements, consequently the briefest review will suffice.

Frames. The boiler is carried on frames forming a *chassis*, which is in turn carried by the wheels (Fig. 5). The frames consist essentially of two longitudinal members connected at the front and back by buffer plates. They are also stayed transversely by the cylinder casting, motion plate, and intermediate cross-stays.

Frames have to be made strong enough to counteract the alternate tensional and compressional stresses set up by the steam acting on each end of the cylinders alternately, and also they have the pull of the engine to transmit to the draw-bar. Locomotive frames are arranged vertically between the wheels and are invariably made of mild steel plates about 1 in. to $1\frac{1}{8}$ in. thick. Owing to the gauge, the distance between them is limited to about 4 ft. 2 in.

They are merely strong plates shaped out to take the axles of the wheels, and suitably drilled for the attachment of all the engine details. They are rolled from Siemens-Martin mild steel ingots. An average frame-plate ingot weighs about five tons, and will make two plates. Frames, previous to about 1868, were made in three lengths welded together, as at that time machinery did not exist for rolling them in one length. In America, Bavaria, Austria, and other countries plate frames have in recent years been displaced by bar frames. As their name implies, these are usually built up of bar sections welded or strongly bolted together and consist generally of two main portions, a front section supporting the cylinders and motion parts, and a back portion which accommodates the axle-box guides or pedestals. Connection between the two sections is made by two arms forming an extension of the front pedestal, between which is spliced, bolted and keyed the front portion of the frame. The length of frame between the leading and trailing wheels is usually doubled, the upper bar or 'top rail' and the lower member or 'bottom rail' being stayed together by ribbed uprights, which are utilized for suspending the brake hangers. Sometimes the extension passing beneath the cylinders is also doubled, but for large engines the two bars give place to a single slab to which the cylinders are bolted. The pedestals are stayed across the openings by pedestal-binders equivalent to horn-stays. The cylinder castings are relied upon to bear the greater portion of the burden of keeping the frames in alignment, the remainder of the staying being obtained by broad, well-ribbed cross-ties running horizontally and diagonally between the frames and placed as close to the pedestals as possible. The frames are usually of wrought iron about 4 inches in section. It is of interest to note that the earliest locomotives were fitted with bar frames : subsequent engines had wood frames plated with iron and a wood buffer beam at each end.

Frame failures are of fairly common occurrence in America, and, being expensive to repair, the practice of using cast steel frames having an $\mathbf{I}$, instead of a rectangular section, is being extensively introduced. From theoretical considerations a frame of this type, allowing the same amount of metal and the same width of frame, has been shewn to be about four times as strong in the horizontal plane and a little over half as strong in the vertical plane. The disadvantage is that welding is difficult if the frame is broken.

Advantage may be taken of this opportunity to state that generally, in the construction of locomotives, steel castings are replacing, more and more, forged pieces. The Belgian State Railway not long since gave builders a long list of parts, at present forged, which they are allowed to replace at will by steel castings. In Hungary foundation rings are steel castings, and in Germany bar frames have also been cast of steel. Other details such as guide-bars, cross-heads, frame-braces, stretchers, smoke-box saddle, buffers, and buffer-brackets, together with brake-beams and a number of lesser parts, have also been made of steel castings. It is, of course, not necessarily a cheap engine that is built of steel castings, but it may be, and generally is, much lighter than if forgings had been used. To resist the strains of traction and of buffing, box-girder frames have also been used on heavy goods engines on the Continent. To eliminate the element of weakness inherent in welded or bolted parts, bar frames cut from the solid are used in some cases. The writer recently saw a set in the process of manufacture at the works of the North British Locomotive Company. They were being made from solid steel slabs weighing 8 tons each. The holes were first drilled, then the slab was ripped up, planed and finished on a shaping machine. The weight of frame finished was only 1 ton 19 cwts.!

A combination of plate and bar methods has also been tried on foreign locomotives, the rear portion being constructed on the plate system.

Wheels. The wheel comprises tyre, wheel-centre, axle and, in a coupled wheel, crank-pin. The tyres are bored out somewhat smaller than the wheel centres, and are shrunk on to them, a usual shrink-age allowance being $\frac{1}{1000}$ of the diameter of the wheel centre. The tread of the tyre is turned up when

the wheel and axle are finished, and is left rough-
turned to assist adhesion. To enable the wheel
centre to be placed in the tyre, the latter is expanded
in a gas furnace and the wheel centre lowered into it.
Wheel centres were at one time forged, but now are
usually cast in steel. Some railways use cast iron for
the wheel centres of mineral engines.

Wheel centres, after being turned and bored, are
pressed on to the wheel seat of the axle by hydraulic
pressure, at from 8 to 12 tons per inch diameter of
axle.

Several methods of securing the tyre on the rim
are in vogue. On the London and North Western, the
outer edge of the tyre is turned outwards so as to
form a recess and lip, a corresponding projection
being formed on the rim, which fits into the recess.
Set screws are screwed into the rim at intervals with
their ends projecting into the rim.

Axles. A straight axle consists of two journals,
two wheel seats and the shank: in a crank axle
the two cranks occupy the place of the shank. The
journals are case-hardened. A small percentage of
chromium is nowadays introduced into axle steel, as
this has been found to toughen it. Crank axles are
either forged from slabs under the hammer-press or
built up. A built-up axle consists of one piece of
axle for each end, a middle piece, four crank cheeks
and two pins; they are machined and keyways cut

previous to assembling. It is usual to shrink the parts together, but the crank cheeks are all keyed to their respective parts of the axle in addition. The oblique arm-crank axle has an extended application on Belgian and some French locomotives. This is a very strong, cranked axle and is often made hollow throughout except in the oblique portion.

Other running part details. Connecting rods are mild steel forgings, planed, bored and slotted on all working faces, i.e. where brasses or cottars fit, but milled up and polished elsewhere, as flaws shew up better on polished surfaces. The small ends are usually brass lined or bushed for the gudgeon pin, and the big ends fitted with some arrangement of adjustable brass.

Coupling rods are bushed at each end for the crank-pin, and machined out to a channel section, as this style combines lightness with strength. Eccentric sheaves and straps are usually of cast iron, and the rods of Yorkshire iron. The straps are often fitted with a removable cast iron liner which can be easily renewed. The sheaves are keyed and also held by pinching screws. All the motion pins, etc., are made from mild steel and are case-hardened and ground up true after machining. Axle boxes are steel castings with a semi-circular brass strip with white metal insets called a 'step' let into the bearing part of the casting for the axle journal to wear against. The edges of the box are planed to wear against the horn block faces. These horn blocks, or axle box guides (cf. Fig. 5), are of cast steel and are fixed to the frames by rivets. A distance piece and a strong bolt through the ends of the horns are fitted to stay the bottoms of the horns of the frame together.

Radial Axle Boxes, Bogies, and Bissels. In the earlier stages of locomotive building it was customary to rely on the flange of the driving wheel for guidance and to force the engine to turn when entering a curve. With low speeds this was sufficient, especially as English engineers kept the track in as nearly perfect condition as possible and made these curves of very large radius. With the increasing length of engines some guiding device became imperative, and the radial axle box, bogie and pony truck or Bissel were developed.

No better example of a radial axle box exists than that designed by the late Mr Webb for his engines on the London and North Western Railway. In this a certain amount of side-play is secured by uniting the two axle boxes in one curved casting which is capable of sliding $1\frac{3}{8}$ ins. to the right or left from its central position, over a guide curved to a corresponding arc of circle. The wheel and casting are brought back to their central position by horizontal springs as the engine leaves the curve and

runs on to the straight. They are, however, being
displaced from their position of leading wheels by the
superior devices known as the bogie and pony truck,
although they are coming into favour with *Atlantic*
and *Pacific* type engines, as trailing wheels, in which
the trailing axle being placed from 9 to 12 ft. behind

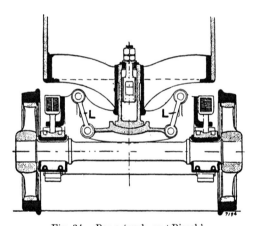

Fig. 34.   Pony truck or ' Bissel.'

the rear coupled axle, must necessarily be allowed
the same 'play.'   The pony or Bissel truck, Fig. 34,
is used when the weight is not too great for one pair
of wheels.   It is pivoted by means of a radius bar to
some point on the frame in rear of its axle, the whole
truck frame being free to slew sideways under the

point of the engine except as controlled by swinging links L, these being intended to allow the turning forces to act gradually on the front part of the engine. Sometimes horizontal springs are used instead of swing links. Their action in this connection has been explained above in dealing with the radial axle. The swinging links are attached at their top ends to the truck frame and at the lower ends to brackets on the socket of the engine centre pin. On entering a curve, the side movement of the truck causes one of the links to shorten in its effective vertical length. This has the effect of lifting the engine or, what is the same thing, putting more pressure on the springs at that side, which pressure tends to bring the truck back to its central position on leaving the curve. The truck is so designed that the engine forces the centre line of the truck to take a direction parallel to the tangent at the point on the curve where the truck wheels are bearing. This is accomplished by pivoting the truck at such a distance from the truck centre pin that the angle θ through which the truck turns, will be greater than the angle between the tangent to the curve where the truck bears and the centre line of the engine. The outer truck wheel flange always bears against the outer rail when θ is less than this angle and against the inner rail when θ is greater. The exception to this rule occurs when the weight on the truck is not sufficient to prevent the truck from being slid against the outside rail.

The now familiar four-wheeled truck or bogie, for carrying the leading end of the engine, was introduced in America, and for a long time was regarded with disfavour by English engineers. The necessity of securing a flexible wheel base with an increasing weight of the leading end of the engine has, however, ultimately led to its extensive adoption. As first introduced it was allowed only a simple rotating movement about its centre pin upon which the front end of the locomotive rested. They were then made with swinging links, as described above, so as to allow a certain amount of swing or horizontal play perpendicular to the centre line of the truck. In some designs the links are replaced by spiral or laminated springs on each side to secure control, but there is some divergence of opinion as to which method is preferable. While swinging links are held to possess the advantage in smoothness of action and freedom from friction they are stated by some not to give enough pressure, and this can only be secured by the use of springs.

The theory of the action of the bogie and the calculation of the flange pressure and stresses set up involve a somewhat complicated mathematical treatment outside the scope of this manual.

Wheel Arrangement. The wheel arrangement is

important inasmuch as, in conjunction with the evaporative capacity of the boiler, it largely determines
the type and function of the locomotive. According
to the old system of classification, engines were divided into four classes, namely, (1) express, (2) mixed,
(3) goods, (4) local or tank. Further, when three
pairs of wheels were coupled, they were known as
'six-coupled' engines; four-coupled' denoted that
two pairs of wheels were connected by coupling rods,
and when one pair alone was used for driving, the
locomotive was designated a 'single' engine. For
many years, express passenger single engines enjoyed
great popularity because they were very free running
and capable of attaining high speeds with a load
suited to their power.

The old and somewhat confused system of classification has now been displaced by a notation which is
at once capable of indicating explicitly any particular
type of engine. In using it one is supposed to be
standing on the footplate of the engine and looking
ahead, and considering in succession the bogie or
leading wheels, driving and coupled wheels, and
trailing wheels. Thus a 'single' engine would be
designated a 2–2–2 type; another, with a leading
bogie and four coupled wheels, would be represented
in the notation as belonging to the 4–4–0 class and
so on, the 0 indicating the absence of any trailing
wheels. A four-wheel coupled-in-front engine with

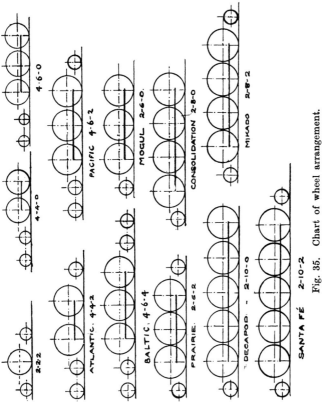

Fig. 35.   Chart of wheel arrangement.

no leading wheels and a trailing bogie would be designated a 0–4–4 type. The principal types in favour to-day are set out in a diagrammatic chart, Fig. 35, which will enable the method to be readily followed.

The prevailing types of the modern express passenger engine in this country are the 4–4–2 *Atlantic* and the 4–6–0 types (see Figs. 18 and 20), although a large number of the older 2–4–0 and 4–4–0 types are still in service. The *Mogul* 2–6–0 type is represented on the Great Western Railway. The 0–6–0 and 0–8–0 are popular types of heavy goods engines, and examples of the 4–8–0 and 0–8–4 engines are represented by heavy shunting engines.

On the whole it may be said that the *Atlantic* type is generally the most popular to-day, chiefly for the reasons that it permits the use of a long boiler barrel with an augmented volume of water, large fire-box and a high evaporative efficiency ; comparatively short coupling rods ; an advantageous distribution of weight ; finally a small fixed-wheel base which renders the engine easy on curves. Although it may mean a diminished stability if the trailing wheels are allowed an excessive uncontrolled radial movement, it stands for the highest degree of development of the four-coupled engine and is *par excellence* the type of modern specialized express engine for all but the heaviest loads and severe gradients.

CHAPTER VIII

STABILITY

THE exterior forces acting on the locomotive as a whole, assuming that the track is perfectly straight and horizontal, are, besides the weight, the effect due to the resistance of the train and the air, the tangential reaction of the rails on the coupled wheels, the forces due to the inertia of the reciprocating parts, and the horizontal component of the centrifugal force of the revolving masses where they have not been completely balanced. Other factors tending to set up perturbations in the movement of the engine are variations of the effort on the piston, and the elasticity of the draw gear.

The disturbances due to the centrifugal forces and centrifugal couples set up by the revolving masses are such as to develop oscillations in the engine unless they are balanced so as to reduce them to the smallest possible amount. A familiar example of centrifugal force occurs when a stone or small bullet is whirled round at the end of a long, fine string. This string itself is pulled away from the centre by the bullet, which is said to exert on it a centrifugal force. A simple extension of this example is the face plate of a lathe to which is eccentrically attached a heavy

piece of work. If this is revolved, a wobbling or
vibrating motion results which sets up stresses in
the frame and bearings varying in magnitude as
the square of the angular velocity, and directly as the
distance between the axis or spindle and the revolving
mass. If continued the bearings would wear rapidly
and unequally, hence when work of this kind has to
be dealt with, it is balanced by attaching a piece
of iron of equal weight to the opposite side of the
face plate. It is not difficult to see from this that
a crank is also an example of an unbalanced force.

To illustrate a centrifugal couple, let a cord be
attached to the centre of a stick and whirled round.
It will be found to always set itself radially to the
axis of revolution, which is due to the centrifugal
couple set up.

A couple, it may be explained, is the name given
to a pair of equal and opposite forces acting in
parallel lines.

Two equal masses, such as two cranks set at 180°
to each other and revolving in different planes on the
same shaft, furnish another example. Each develops
a centrifugal force equal in magnitude and opposite
in sign, and form a couple tending to turn the shaft in
the plane of the axis of revolution. But it is not only
the centrifugal forces set up by the rotating cranks
that require to be balanced, there are also to be
considered the disturbing effects of the reciprocating

parts such as the piston, piston rod and crosshead;
also those due to the motion, varying from a straight
line to a circle, of the connecting rod.

Briefly, the principle by which such a system is
established more or less in equilibrium is, to quote
Rankine, 'to conceive the mass of the piston, piston
rods and connecting rods, and a weight having the
same statical moment as the crank, as concentrated
at the crank pins and to insert between the spokes of
the driving wheels counterpoises.' The weights and
position of these balancing masses can be approximately determined by calculation. Balancing the
revolving masses, such as crank-arms, crank-pins,
etc., offers a problem of no special difficulty. It is
accomplished either by prolonging the crank-arms
to the opposite side of the axle to form a balance
weight, or by putting balance-weights on the inside
of the wheel rims. The first method avoids the
centrifugal couple between the wheels, which sets up
a bending moment on the axle. The last-named
method is the most favoured. It was used as far
back as 1842, when MacConnell, at the suggestion of
Heaton, a Birmingham engineer, adopted balance
weights on the Birmingham and Gloucester Railway.

The problem of balancing the reciprocating parts
consisting of the piston, piston rod and crosshead,
and that portion of the connecting rod which reciprocates, is not so simple. For a full discussion of

the question the reader is referred to Prof. Dalby's standard treatise on the subject. In this place we can only summarize some of his results.

Where

M is the mass in pounds of the reciprocating parts belonging to each cylinder;

r the crank radius in feet;

n the revolutions of the crank axle per second;

d the distance in feet between the cylinder centre lines;

t the distance between the centre line of the driving axle and the line of traction.

The unbalanced reciprocating parts cause

(1) An unbalanced force, the maximum value of which is given by $1\cdot7\,Mn^2r$ lbs. weight. This force accelerates the whole mass of the train positively and negatively in the direction of travelling. The considerable efforts set up by this couple react upon the draw-gear of the locomotive, and give rise to a jerky motion which severely tests the strength of the engine frames.

(2) A couple whose maximum value is

$$0\cdot85\,Mn^2rd \text{ foot-lbs.}$$

This couple produces an oscillatory motion about a vertical axis, which, superposed upon the general forward motion of the engine, causes a swaying from side to side. This, when acting on a short engine, may become dangerous at high speeds.

(3)   A couple whose maximum value is

$1{\cdot}7\,Mn^2rt$ foot-lbs.

This couple tends to cause oscillation in a vertical plane about a horizontal axis.   The reader can work out the values for himself from the following data for weights: connecting rods, reciprocating portion, 180 lbs.; piston, 150 lbs.; piston rod, crosshead, and pin cottar and nut, 180 lbs.

The best method of balancing reciprocating parts is not immediately obvious ; fortunately, however, it can be shewn that although they move in straight lines or describe ovals in a vertical plane, it comes to the same thing if they are considered as a body of equal mass revolving in a circle whose centre is the crank axle.   Therefore in a two-cylinder engine the reciprocating force and couple are dealt with in the same way as those due to the revolving masses, namely, by placing balance weights inside the rim of the driving wheel.   Separate weights are not used for each set of parts, but are combined.

Unfortunately in curing one trouble another is created, for the revolving balance-weights themselves set up a centrifugal force, which acts at right angles to the plane in which the reciprocating parts move. The horizontal component of the force tends to thrust the axle box against the guides, and the vertical component acts by lifting the wheel at one instant and dropping it, thus causing on the one hand slipping, and on the other a 'hammer blow' action on the rail.

This hammer blow can amount to as much as 25 per cent. of the static weight, and is not only injurious to rails, tyres and bridges, but limits the carrying capacity of the axle. The locomotive engineer is thus on the horns of a dilemma : either he can fully balance the reciprocating parts and so eliminate the swaying couple ; or he can leave them unbalanced, doing away with hammer blows and leaving the engine to lurch along the road with the certainty of derailment if very high speeds are reached.

Practice has regulated the amount of compromise, which is obviously the only way out, by balancing completely the revolving parts and only three-fourths of the weight of the reciprocating parts.

The connecting rod, which partly revolves and partly reciprocates, cannot be perfectly balanced by a rotating weight, but can be balanced in a vertical direction by dividing the masses of the rod between the crank-pin and crosshead. The obliquity of the connecting rod is also responsible for a variation in rail load which, compared with the effect of the balance weights, may, however, become negligible at high speeds.

The force acting at the crank is accompanied by a reaction on the slide bar, and forms a couple tending to move the frame in a direction in which the forces act and, as the locomotive is spring borne, to vary the load on the springs. Its magnitude depends upon the speed, the amount cut-off in the cylinder and the strength of the springs, and the relative length of the rod and crank.

The problem of balancing is more readily solved with four-cylinder engines in which, by setting the cranks at 180°, the reciprocating parts are automatically placed in equilibrium and need no balance weights. This disposition, however, introduces costly complications by the fact that each cylinder requires the use of an independent valve gear. A practical way out of the difficulty is secured by arranging the cranks in pairs at 180°, the pairs themselves being at 90°. The reciprocating masses are then balanced except for a horizontal swaying couple which is inconsiderable. An excellent example of an engine balanced in this manner is furnished by Mr Hughes' six-wheel coupled, four-cylinder passenger engine on the Lancashire and Yorkshire Railway. The outside reciprocating parts weigh, roughly, 595 lbs., and the inside parts 524 lbs., so that they are for all practical purposes balanced by the cranks being set as mentioned above. The outside revolving masses total 1108 lbs. and those inside 1255 lbs. These are balanced by the method already mentioned of prolonging the crank-arms to form balance weights. The connecting rod is also dealt with as stated earlier. There remains to be balanced, the unbalanced portion of the crank boss, part of the crank-pin and coupling rod, which are dealt with by balance weights placed in the leading and trailing parts of driving wheels.

The tangential reaction of the rails on the driving wheels involves a consideration of the torsion set up in the driving axle. If the engine has outside cylinders, the effort set up on the right hand crank pin, for example, can only be transmitted to the left hand wheel by a torsional stress in the axle, and whatever its magnitude, this implies a corresponding backward slipping of the right hand wheel, which, however, cannot take place so long as the limit of adhesion has not been passed. In the case of an inside cylinder engine, the effort on each piston is transmitted to the wheels in inverse proportion to the distance between its centre line and the wheel. The effort on the piston also sets up an effort, call it P, in an opposite sense on the cylinder-fastening and therefore on the main frame : also the effort set up by the crank sets up a corresponding reaction in the axle boxes p, so that the real effort exercised by the engine is the difference $P - p$. Alternating stresses thus occur in the frames between the cylinders and the crank-axle which may be sufficiently high to cause weakening of the cylinder attachments and sometimes rupture of the frames. Another effect is

to set up a pressure, first in one direction and then
in another, between the journals and bearings, and
between the axle boxes and their guides which
causes a 'knocking' when there is any play between
the parts.

All who are connected with railways know
that certain types of locomotives when driven at
high speeds, oscillate vertically and laterally to a
dangerous extent even on a good road. Locomotives
coupled in front with a bogie under the footplate
(0–4–4 type) are held by some to be specially undesirable for high speed running. An example of this was
furnished in 1895 by the Doublebois accident on the
Gt Western Railway, in which two such locomotives
hauling a fast passenger train, left a practically new
road, with easy curves. Single engines with a single
pair of leading wheels develop a considerable swaying and plunging motion which the introduction of
the bogie, however, largely removed.

There remains to examine the load variation on
the wheels and springs set up during running by
imperfections of the road, and which may increase
to a very considerable extent any tendency to swaying or side motion of the engine set up by its own
action.

An oscillatory or galloping movement about a
transverse axis of the engine is produced by the
rail joints or inequalities in the surface of the rail.

Readers will doubtless have noticed during the passage of a train a depression at the joints which results in the rail presenting to the passage of a locomotive a curved form, the lowest point of which is at the joint. The difference of level in the rail itself, according to M. Coüard may reach 4 mm. on good, and 8 to 10 mm. on imperfect roads. The oscillations so set up do not reach a dangerous proportion, except when, by being added to each other, their amplitude goes on increasing. This increase takes place if the interior friction of the springs does not suffice to deaden the oscillations, and the period becomes that due to the disturbing force, namely, the length of a rail. The use of compensation levers between the springs—a standard practice in France and America—probably has a considerable influence on this type of oscillation.

Oscillations about a longitudinal axis situated in the plane of the crank axle are set up by difference in the level of the two rails. They become permanent when the joints are set one in front of the other, but are damped out by the interior friction of the springs. When the joints are directly opposite to each other, similar oscillations are also set up at non-symmetrical points in the rails, at crossovers, and under certain circumstances at the entrance to a curve. In the latter case they are due to the application of centrifugal force. This is a convenient place wherein to examine

the conditions necessary for an engine satisfactorily
to negotiate a curve.

A very considerable effort is required to keep an
engine on the rails and prevent it persisting in a
straight course according to Newton's first law of
motion. This effort is given by the formula for
centrifugal force acting in a horizontal direction at
the centre of gravity of the engine, as follows :—

$$F = \frac{P \times V^2}{g \times R},$$

where

$\dfrac{P}{g}$ = the mass, i.e. weight divided by the force of
　　　gravity,

V = the velocity in feet per second,

R = the radius of the curve.

In early locomotives it was considered important
to keep the centre of gravity low, but more recently
it has been seen that a high centre of gravity, within
certain limits, contributes to steadiness of running
and diminishes the lateral pressure upon the outer
rail in curves, by reducing the obliquity of the line
of thrust on the rail and tending to make it more
vertical. The load on the outside wheel is increased
and, therefore, the resistance to derailment. The
height of the boiler has been increased from 5 ft. 3 ins.
in early engines, to 7 ft. 11 ins. and more to allow
larger boilers to be used with large diameter driving

wheels. It is important to notice that as the boiler forms less than half the total weight of the locomotive, the centre of gravity of the whole engine is only raised by less than half the amount the boiler is raised. Beyond a certain limit, however, raising the centre of gravity increases the liability of the engine to overturn by increasing the overturning moment due to centrifugal force.

Two simple calculations based on actual data derived from the American boat train disaster at Salisbury Station in 1906, will suffice to illustrate the approximate condition of stability of a modern locomotive at speeds of 30 miles and 70 miles per hour respectively. The following are the necessary data :—

(W) Weight of locomotive ...... 54 tons,

(V) Velocity (30 miles an hour) 44 ft. per sec.

(V) Velocity (70 miles an hour) 102·66 ft. per sec.

(g) Force of gravity .... 32·2 ft. per sec.

(R) Radius of the curve at Salisbury .... 523 ft.

(h) Height of centre of gravity of engine above rail level 58·5 ins.

(l) Horizontal distance between vertical line through centre of gravity of engine and outer rail, allowing for 3½ in. super-elevation 31·75 ins.

The speeds of the train just before reaching Salisbury were worked out by Mr Holmes, the superintendent of the line on the L. and S. W. R. as follows:—

Section of Line	Length in Miles	Speed in Miles per hour
Tisbury to Dinton	4·29	64·3
Dinton to Wilton	5·83	69·9
Wilton to Salisbury (West) ...	2·29	68·5

It should be pointed out that for about half the distance beyond Wilton there is an up gradient naturally causing diminution of speed, while the remainder of the section is on a down gradient of 1 in 115, so that to account for the average speed of 68·5 miles an hour the train probably travelled at the rate of at least 70 miles an hour between the west box and the point where the accident occurred.

The moment of the weight of the locomotive about the outer rail is obviously

$$W \times l = 54 \times 31{\cdot}75 = 1714{\cdot}5 \text{ inch-tons},$$

and so long as this exceeds the overturning moment due to centrifugal force, the stability of the engine will be assured, but of course it may be in a critical condition in this respect unless an ample margin of safety exists.

At the velocity of 30 miles an hour, and using

the formula given above, the centrifugal force (f) developed is

$$f = \frac{W \times r^2}{g \times R} = \frac{54 \times 44^2}{32\cdot2 \times 528} = 6\cdot15 \text{ tons,}$$

and the moment of the force about the outer rail is

$$f \times h = 6\cdot15 \times 58\cdot5 = 359\cdot77, \text{ say 360 inch-tons.}$$

Comparing this value with that of the moment of stability of the engine, it is evident that at the speed of 30 miles an hour, a very ample factor of safety exists.

At the velocity of 70 miles an hour, the centrifugal force developed is

$$f = \frac{54 \times 102\cdot66^2}{32\cdot2 \times 528} = 33\cdot5 \text{ tons,}$$

and the moment of the force about the outer rail is

$$f \times h \; 33\cdot5 \times 58\cdot5 = 1960 \text{ inch-tons,}$$

which, being in excess of the moment of stability of the engine, shews that at the speed of 70 miles an hour round the curve in question, it is impossible for the engine to avoid being overthrown.

By calculations similar to the above it would be easy to demonstrate that the velocity giving rise to an overturning moment exactly equal in value to that of the moment of stability for the engine and curve here considered is equal to the speed of about 66·2 miles an hour.

To add to the safety of engines in running round curves, super-elevation of the outer rail is resorted to, the amount being given by the formula

$$E = G \; \frac{V^2}{1 \cdot 25R},$$

where

$G =$ Gauge in feet,

$V =$ Velocity in miles per hour,

$R =$ Radius of the curve in feet,

$E =$ Super-elevation in inches.

On the Salisbury curve the super-elevation was $3\frac{1}{2}$ in. and therefore suited to a speed of only 34 miles per hour, and according to the working time-sheets the speed over the curve for non-stopping trains should not have exceeded 25 miles per hour. The effect of super-elevation is to throw the centrifugal component, acting downwards, well within the gauge. If it falls outside, the engine is unstable.

There is yet one other disturbing effect to be noticed, that is a jerking movement from side to side set up by defects in the road, the coning of the tyres and the flexibility of the suspension system. Suppose the engine to be deviated to the left a lateral force is set up, causing abrupt contact between the flange and rail, and resulting in a more or less violent shock. If the rail is sufficiently elastic the force will be completely neutralized, if not it

will be more or less completely given back to the
wheels, which will result in an impulse towards the
opposite rail. Assisted by the coned shape of the
tyres, a sinuous movement is thus set up which,
unless any further disturbance is set up, will gradually
lessen until the action of the rails restores normal
running.

Axle play, wear of the tyres, coupling arrangements, a swaying couple and other factors modify
the circumstances of this movement. When axles
are given an amount of play in relation to the main
frames, they may take a sinuous movement independent to that of the engine itself. For example,
in the case of a bogie it has been found that if the
lateral control is insufficient, it will float along or
'get across' the road. The bogie as well as the
Bissel however possesses the great advantage of
attenuating shocks and so relieves the stress on the
permanent way. Although their mass is relatively
small and their lateral reaction equal only to the
tension of the controlling springs or pressure on
the swinging links, they are capable of exerting a
powerful leverage tending to straighten out the
running of the engine.

One more feature of the bogie may be mentioned,
that is, its tendency to flatten the road for the driving
wheels. When a wheel passes over a particular point
in a rail, that point, owing to the elasticity of the

rail, is depressed, and as it is depressed the upward pressure of the rail is increased. As the pressure on the rail is removed, the upward pressure diminishes, but not so quickly as it had increased. In other words there is a 'lag' effect. Thus the wheels of the bogie carrying the weight of the front portion of the engine would set up a deflection in the rail which, owing to the lag, the driving wheels would not have to repeat. The late Mr Webb, in a discussion at the Institution of Civil Engineers, stated that he well recollected Mr Patrick Stirling saying one day, when he had been discussing with him how he had managed to pull the heavy trains on the Great Northern Railway with single driving wheels, 'Mon, I have the weight on the bogie, and it lays the road down for the single wheel to get hold of it.'

CHAPTER IX

PERFORMANCE AND SPEEDS

THE accurate determination of the factors which enter into the efficiency of the locomotive, although of the utmost importance in practice, for a long time remained a matter of rule of thumb. Rough and ready running tests no doubt contributed a great deal of valuable information to the locomotive engineer, but it was not until quite recent years that the scientific precision adopted in stationary engine testing came to be applied. With the advent of the dynamometer car and experimental testing plants means have been placed at the disposal of the locomotive engineer for determining such points as the average rate of fuel consumption, steam consumption at various speeds, the average indicated horse-power, the effect of various cut-offs as shewn by indicator diagrams, smoke gas analysis, resistance, drawbar pull, etc.

Such information may be ascertained in two ways, first, under actual service conditions on the road, using a dynamometer car; second, under laboratory conditions, with the engine stationary and the use of a testing plant. The latter method gives very exact results which, however, are of value only in so far as they are tested by actual running experiments.

For conducting the latter the engine is fitted up to enable indicator diagrams to be taken, and a dynamometer car is included in the train between the tender and first vehicle. The use of the latter will be understood by briefly describing the equipment of such a car on the North Eastern Railway. The body of the car is built on a steel underframe shaped to take a special spring consisting of thirty steel plates, and each separated by rollers to minimise friction. From the spring the pull is transmitted to the train. As the spring is deflected it moves a stylograph pen over a roll of paper, thus producing a curve of drawbar pull. The paper is caused to travel over a table by drums driven from a measuring wheel which rolls on the rail, thus enabling the operator to determine the speed of the train at any instant. The permanent speed record is given by a pen in electrical communication with a clock, which makes a mark on the travelling roll of paper at two-second intervals. The speed can be read off from this by the aid of a special scale. Dials shew the distance travelled. A boiler pressure recorder is also fitted, and a meter is constructed on the same principle as a planimeter for registering the work done. In this apparatus, a horizontal circular plate moves a proportional distance to that of the train, whilst a frame supporting a small wheel on edge moves across it from the centre a distance proportional to the pull on the drawbar. Its revolutions are therefore a measure of the work done, and as it is in electrical communication with a meter, the work is recorded. An indicator records the pressure in the steam chest, and another registers the velocity of the wind, which blows down a tube kept facing its direction, and causes the rise and fall of a pen on the paper drum. The direction of the wind is indicated by means of a dial in the roof.

Most British railways have dynamometer cars,

but the testing plant possessed by them for testing locomotives under laboratory conditions is not conspicuous for its high value and consists chiefly of friction rollers. In 1890 the Purdue University, Indiana, put down the first plant permitting tests to be carried out on really scientific lines whose example was followed by the Pennsylvania Railroad and later by the Swindon Works of the Great Western Railway. With these plants the locomotive to be tested is placed on a system of rollers whose centre lines are directly underneath those of the locomotive axles. It is kept in this position of unstable equilibrium by a very ingenious elastic coupling, which measures the tractive effort of the locomotive at its wheels, while, at the same time, it prevents any displacement which could endanger its equilibrium. The rolling resistance which results, when running, from the resistance proper of the locomotive and that of the train which is being hauled, is produced by a hydraulic brake acting on the supporting rollers: in consequence of this action, these oppose to the rotation of the driving axles a resistance similar to the reaction of the rail. A revolution counter shews at every moment the speed the locomotive would have if running on an ordinary track. It is easy to see, how, under these conditions it is possible to make certain experiments which it would be very difficult to carry out while running on the track, such as, for instance, the measurement of the amount of coal which can be burnt per square foot of grate and per hour, the consumption of steam at different speeds, and for different settings of the valve gear. As previously stated it has the disadvantage of placing the locomotive under artificial conditions, and the most serious defect is the impossibility of studying two important elements, namely the adhesion and the rolling resistance of the locomotive. As examples of some of the results obtained with the Pennsylvania and Purdue plants which may probably be safely applied, the following out of a large number will be useful for reference:

1.   When working at maximum power the boilers tested generated 12 pounds of steam per square foot of heating surface per hour.

2.   The evaporative efficiency falls as the rate of evaporation increases.   When working at full power between 6 and 8 pounds of water per pound of dry coal were evaporated.

3.   Fire-box temperatures according to the rate of combustion range from 1400° F. up to 2300° F. and smoke-box temperatures from 500° F. up to 700° F.

4.   One indicated horse-power per hour was developed with a steam consumption of from 23·8 to 29 lbs. in a simple, and from 18·6 to 27 in a compound engine.   It varies with the speed and cut-off.

5.   A steam locomotive can deliver one horse-

power at the drawbar on a consumption roughly of about 2 lbs. of coal.

6. The mean effective pressure varies with the speed and cut-off, e.g. speed 25 m. p. h. cut-off 6 in., 8 in. and 10 in., the mean effective pressure was 30·5, 51·2 and 63·3 lbs. per sq. in. respectively. At 35 m. p. h. on the same cut-offs, the m. e. p. was 29·6, 42·4, and 48 lbs. per sq. in., the boiler pressure being 130 lbs.

A few typical results obtained from tests actually made on the road may now be given. They will be more conveniently stated in tabular form.

A series of coal consumption observations made on the London and North Western Railway between shallow fire-box engines of the *Experiment* class and deep fire-box engines of the *Precursor* class, both classes working the heaviest and fastest trains between Euston and Crewe under identical conditions, gave the following results :

The *Precursor* engine ran 34,348 miles, and burnt 882 tons 5 cwts. of coal—equivalent to an average consumption of 57·53 lbs. per mile.

The *Experiment* engine ran 34,013 miles, burning 793 tons 8 cwts. of coal, the equivalent average consumption being 52·25 per mile.

Speeds. The performance of a locomotive is generally associated in the mind of the average traveller with speed, without regard to any other

DETAILS OF LOCOMOTIVE TESTS.

Particulars	L. & S.W.R. Express Engine	Lancashire and Yorkshire Simple	Lancashire and Yorkshire Compound	M. R. Express	L. & N.W.R. Express Compound
Class of Engine	4-coupled	8-coupled	8-coupled	Single	4-cylinder express
Diameter of Driving Wheels	7 ft. 1 in.	4 ft. 6 in.	4 ft. 6 in.	7 ft. 9 in.	7 ft. 1 in.
,, ,, Cylinders	19 in.	20 in.	(4) 15½ and 22 in.	19½ in.	(4) 15 and 10½ in.
Stroke	26 in.	26 in.	26 in.	26 in.	24 in.
Average of Boiler Pressure (lbs. per sq. inch)	167·5	178	180	—	200
Grate Area (sq. ft.)	18·14	23·05	23·05	—	20·5
Coal burnt per sq. ft. of grate area per hour (lbs.)	62·54	111·5	74·3	—	—
Coal per I.H.P. per hour (lbs.)	2·31	3·7	3·1	2·9 to 3·1	—
Water evaporated per lb. of coal (lbs.)	9·23	—	—	—	7·8
do. do. from and at 212° (lbs.)	11·35	—	—	—	—
Feed Temperature	61° F.	—	—	—	—
Steam consumption per I.H.P. per hour (lbs.)	—	23·3	18·0	—	—
Weight of train (tons)	—	583	585	—	329 tons 5 cwts. (max.)
Average drawbar pull (tons)	—	4·37	3·79	—	5·25
Average speed (miles p. h.)	—	21·9	23·3	—	50
Indicated H.P.	490·6	701	546	400	—
Coal Consumption in lbs. per mile	—	—	—	—	43·7 lbs.

consideration. There exists too a popular vague idea that electricity is to make practicable hitherto unimaginable travelling speeds. This idea is difficult to account for on any other basis than that electricity is capable of doing for us everything that has hitherto been found impossible without its aid. No high travelling speeds have been attempted commercially, and the Berlin-Zossen high-speed tests in which a specially built racing electric motor vehicle ran the distance of 14·3 miles in 8 minutes, attaining a maximum speed of 210 kms. (130 miles) per hour, have no smack of commercial economy about them. This experimental car hauled no load, and if as much money, trouble and time had been spent upon a high-speed steam racing machine, we should probably have learnt that the same velocity is possible with steam locomotion. The only condition favourable to high-speed and long-distance electric traction is to run single or two coach flying expresses at small time intervals, involving clearing the line of all other traffic. Such a condition for the generality of our main lines and having regard to the standard of comfort, such as dining, sleeping and heavy baggage accommodation, required by passengers to-day, puts the method out of court in favour of heavy express trains in the haulage of which the steam locomotive has the advantage. The commercial aspect of this question is much too strong a factor in railway

administration ever to be sacrificed to any consideration which would involve the most serious of outlays
and the most doubtful of returns.

The attainment of high speed is by no means confined to electrically propelled vehicles. The highest
speed ever authentically recorded in favour of the
steam locomotive is given as having been reached by
the high-speed engine constructed by the Prussian
State Railway Department and exhibited at the
St Louis Exhibition. In the course of its trials, this
locomotive maintained a speed of 82 miles an hour
with a six-car train, representing a tonnage of 240;
a speed of 87 miles an hour with five cars, 200 tons;
and a speed of 92 miles with three cars, 120 tons.

In this country it is doubtful if any higher speed
has been authentically recorded under modern conditions, i.e. hauling a passenger train of average
weight, than that reached on the Gt Western Railway
by the No. 1 *Ocean* up special express on August 30,
1909. The occasion was the opening of the new
harbour at Fishguard, and the Cunard steamship
Mauretania having beaten her previous best time
from New York in a passage of 4 days 4 hours
27 minutes, it was probable that the Gt Western
Railway would also try for a record. The writer
was on the train and recorded a maximum speed
of 90 miles per hour, which was verified by the
representative of *Engineering*, who was also a

passenger. A portion of the run of this train is
plotted on the accompanying chart (Fig. 36) against
the gradient profile: The engine taking the train
from Cardiff to Paddington was the *King Edward*,
the first of a batch of four-cylinder six-coupled non-
compound locomotives now hauling the fastest trains
of the company. The load consisted of 10 eight-
wheeled bogie carriages aggregating 274 tons, and
the running time between Cardiff and Paddington—
a distance of $145\frac{1}{4}$ miles—was $141\frac{1}{4}$ minutes, giving
an average speed of 61·6 miles per hour. This average
speed over a long distance, although magnificent, was
surpassed in the race to Aberdeen, but it is doubtful
if the maximum, 90 miles per hour, was exceeded on
that occasion. This occurred in running down the
1 in 300 bank between Badminton and Wootton
Bassett, although incidentally an equally meritorious
performance was the $23\frac{1}{2}$ miles up hill from Severn
Tunnel Junction to Badminton, with length of 1 in 100,
1 in 68, 1 in 90 up and ten miles of 1 in 300 up in
29 min. 55 secs. or at the rate of 47 miles per hour.

The highest average speed of a regular train was
attained in the famous race to Aberdeen in August,
1895. This contest arose between the East Coast
partnership, viz. the London and North Western and
Caledonian Railways and the East Coast Companies,
the Gt Northern, North Eastern and North British,
and gave rise to some very fine running. The distance

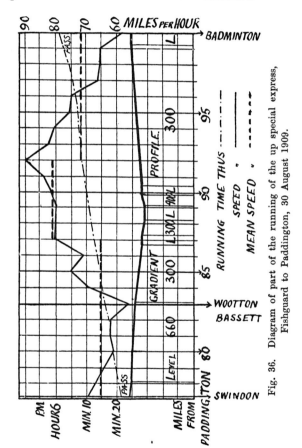

Fig. 36. Diagram of part of the running of the up special express,
Fishguard to Paddington, 30 August 1909.

from Euston to Aberdeen is 539¾ miles and from
King's Cross to Aberdeen 523½—nearly 17 miles
less. In the first case there are the severe climbs
up Shap Fell and the Beattock Summit to be
reckoned with, and on the East Coast the slacks
necessitated by the Forth and Tay Bridges. The
following were the results:—

Date	West Coast		East Coast	
1895	Depart Euston	Arrive Aberdeen	Depart King's Cross	Arrive Aberdeen
Aug. 19	8 P.M.	5·15 A.M.	8 P.M.	5·31 A.M.
,, 20	,,	4·58 ,,	,,	5·11 ,,
,, 21	,,	4·45 ,,	,,	4·38 ,,
,, 22	,,	4·32 ,,	Race abandoned	

The decisive victory gained by the West Coast
partnership on the 22nd involved running at an
average speed for the whole distance of 63·3 miles
per hour including stops, or 64·1 without. The length
between Crewe and Carlisle was run at an average
speed of no less than 65·1 miles per hour. The load,
it is true, was a light one consisting of 3 bogie coaches
totalling 70 tons, but with the type of engine used it
would manifestly have been impossible to have accomplished such an achievement with two or three
hundred tons behind the tender. The engines employed during the race by the West Coast were the

three-cylinder 7 ft. compounds *Coptic* and *Adriatic*
from Euston to Crewe and thence to Carlisle, the older
type *Precedent* class, *Hardwicke* and *Queen*. The
Caledonian Co. used their four-coupled, 6 ft. 6 in. type.
The following are the details of the famous run of
the West Coast on the last day of the race:

	Time		Miles	Engine	Average Speed
Euston	*dep.*	8 P.M.		Webb, 7 ft.	
Crewe	*arr.*	10.28	158	3-cylinder	60
	dep.	10.30		compound	
Preston	*pass*	11.16		*Hardwicke*	
Carlisle	*arr.*	12.36	141¼	*Precedent* class	65·1
	dep.	12.38½			
Perth	*arr.*	3.7½	151¼	Caledonian, 6 ft. 6 in.	60·9
	dep.	3.9½		4-coupled, No. 90	
Aberdeen	*arr.*	4.30		6 ft. 6 in.	
(Ticket Platform			190	4-coupled	65
Station)	...	4.32		No. 17	

It may here be recalled that on Sept. 8th of the
same year, the three-cylinder compound *Ionic* ran
from Euston to Carlisle without a stop—299¼ miles
in 5 hours 53 minutes, i.e. at an average speed of
51 miles per hour, with a load of 150 tons 14 cwts.

The race to Edinburgh in 1888 between the same
companies caused some sensation, and at the time
the performances were unequalled in railway history.

A feature of the contest was that the West Coast train was run as far as Crewe with old single engines of the *Lady of the Lake* class (see page 8) and beyond with 4-coupled *Precedents*. The Caledonian also used a 7 ft. single engine, Mr Drummond's famous No. 123. Mr Stirling's superb singles were used by the Gt Northern Co. as far as York. The best single performance by the West Coast train was on August 14th, when the 400 miles were covered in 427 minutes of running time, or at the rate of 56½ miles per hour throughout; that by the East Coast train was on August 31st, when the 392¾ miles were covered in 412 minutes, a speed of more than 57 miles per hour. The best single performances by each route are given by Mr Acworth as follows:—

Date	Section	Distance (miles)	Time (minutes)	Remarks
Aug. 13	Euston—Crewe	158⅛	166	—
,, 7	Crewe—Preston	51	50	—
,, 7	Preston—Carlisle	90	90	Over Shap summit
,, 9	Carlisle—Edinburgh	100¾	103	Over the Beattock
,, 25	King's Cross—Grantham	105¼	106	—
,, 24	Grantham—York	82¾	88	—
,, 29	York—Newcastle	80¼	81	—
,, 14	Newcastle—Edinburgh	124½	125	Two engines on. Best time with one, 130 minutes, Aug. 31

Every-day speeds in 1911 were of a very high average order. A recent inquiry conducted by the *International Railway Congress Association* brought to light the following facts, as to railways which were in the habit of using speeds of 100 km. (62 miles) in regular service.

In Belgium the speed may attain 110 kilometres (68·4 miles) per hour in cases of delay, on the flat and on down gradients of 5 to 6 per mil., and on curves having a diameter of at least 900 metres (45 chains). The Baden State Railway uses speeds attaining 110 kilometres (68·4 miles) per hour on down gradients not steeper than 4·5 per mil., and on curves having a radius of not less than 1100 metres (55 chains) ; in Germany, this radius is legally pre-scribed as the minimum for a speed of 110 kilometres (68·4 miles) per hour. On several French railways, speeds of 110 to 115 kilometres (68·4 to 71·5 miles) per hour are attained in cases of delay, on down gradients of not more than 5 per mil., and on curves having a radius of at least 700 metres (35 chains); the railways in question are the Midi, the Paris-Lyons-Mediterranean and the French State. Of the railways regularly using speeds of over 100 km. (62 miles) per hour, France is represented by speeds which may attain 120 kilometres (74·5 miles) per hour. This is the maximum speed fixed by the supreme authorities. It is attained on up gradients

of not more than 3·3 per mil., on down gradients of
not more than 5 per mil., and on curves. With the
Bavarian locomotive speeds of up to 150 kilometres
(93 miles) per hour are stated to have been attained.

In connection with the subject of speed it is
noteworthy that the highest velocities have not
been obtained with the largest size driving wheels.
Mr Marshall, in a paper read before the Institution
of Civil Engineers, drew attention to a run in which
11 successive miles were covered by a train drawn
by an engine with driving wheels of 5 ft. 8 in.
diameter. If driving wheels of 7 ft. 6 in. diameter,
with proportionally larger cylinders, could be worked
at the same number of reciprocations per second (13),
a speed of 112 miles per hour would be obtained.
Mr Marshall suggested that the cause of the practical
limitation in the speed of locomotives is not, as
has been generally assumed, the steam not being
able to escape quickly enough from the cylinders,
but should be looked for in a slipping of the driving
wheels, arising from the effective adhesion weight
being seriously reduced, when running at high speeds,
by the vertical action of the disturbing forces of
the balancing masses in the rotation of the wheels.
It has been found that at a speed of 74 miles per
hour, a slip of as much as 19 per cent. takes place in
a four-coupled engine.

The idea put forward that the main requisites

for obtaining high speed are an increase in the number of coupled wheels and a corresponding increase in the boiler power has certainly been borne out in recent practice, and is well evidenced in the example given above of the run of the *Ocean* express on the Great Western Railway.

CHAPTER X

COMPOUNDING

IT was seen in a previous chapter that one of the main causes of loss in the working of an ordinary engine was due to condensation which is set up when the steam at boiler pressure is brought into contact with the walls of a cylinder cooled down by the low temperature of the exhaust steam. The temperature difference may be as much as 140° F., as for example when the working pressure is 180 lbs., with a temperature equivalent of 380° F., and the terminal pressure 10 lbs., with a temperature of 240° F.

High piston speeds may reduce this range. According to Mr Hughes, when the piston speed exceeds 600 ft. per minute, the period is too short to render the difference of temperature due to the interchange of heat noticeable in simple and compound working.

One remedy applied is, as has been seen, to super-heat the steam before its arrival at the cylinder; another and older method is compounding. It is not however with this sole object in view that compounding is resorted to; its adoption means also an efficiency obtained by expanding the steam through more stages than is possible in a single cylinder.

It can be shewn that the efficiency of a perfect heat engine may be measured by the ratio

$$E = \frac{\tau_1 - \tau_2}{\tau_1},$$

where τ_1 is the absolute maximum temperature of the working fluid in the engine and τ_2 the absolute maximum temperature of the working fluid in the engine. The cycle cannot be realised in practice but it indicates theoretically and practice confirms, that the greater the difference between τ_1 and τ_2 the greater will be the efficiency. The largest difference is obtained by compounding. The steam is admitted into one cylinder called the 'high-pressure' cylinder, and expansion allowed to commence therein, and afterwards it is exhausted into a second or 'low-pressure' cylinder, where the expansion is continued.

The number of times the steam is expanded is called the ratio of expansion: e.g. if it is expanded twice, the ratio of expansion is 2 to 1.

It is a convenience in connection with the calculation of horse-power and mean effective pressure to consider the total expansion as referred to the low-pressure cylinder. It is immaterial, for this purpose, what the ratio of expansion may be in each cylinder; it is as though the whole range takes place in the low-pressure cylinder. If the capacity of the latter is, say 9 cubic feet, and steam is cut off after one cubic foot has been admitted to the high-pressure cylinder, then the ratio of expansion is 9 to 1. It is of importance that the work done in each cylinder should be theoretically approximately equal. The exact ratio of the volume of the high- and low-pressure cylinders is, however, a somewhat debateable point, as it ranges from 1 to 1·69 up to 1 to 3 and more.

Owing to the range of temperature and the great differences of piston effort at the beginning and end of the stroke the practical limit of expansion in a single cylinder is about three times, whereas, by further utilizing the steam in a second cylinder, instead of exhausting it to the atmosphere, practically double the amount can be obtained, the limitation being the requirement of a certain amount of pressure in the exhaust steam to serve for the blast.

In engines of the stationary and marine type expansion is continued until the exhaust pressure becomes, by the employment of a condenser, that due to a vacuum, but the limitations of the locomotive prohibit the use of such an apparatus. Nevertheless good results are obtained in stationary engine practice

with non-condensing compound engines, hence it was
thought that the method had only to be applied to
locomotives to secure its general adoption. This may
be said to be the case in the country where it was
first applied, viz. France, and more or less generally
on the continent.

It was first applied by Mallet in 1878 on the
Bayonne and Biarritz railway, with one small high-
pressure cylinder and one large low-pressure cylinder.
Mallet's method was tried with slight improvements
in 1880 by von Borries on the Hanoverian State
railways, but the first to put compound locomotives
into regular use was the late Mr Francis Webb of the
London and North Western Railway. In his earliest
engine two outside high-pressure cylinders were used,
14 in. diameter by 24 in. stroke, which drove outside
cranks on the trailing wheels ; and one large low-
pressure cylinder, 30 in. in diameter and 24 in. stroke,
placed inside the frames and below the smoke-box,
driving on to a crank in the axle of the leading pair
of driving wheels. Thus only one, and that a very
large cylinder, exhausted to the chimney which gave
these engines a characteristic 'beat.' A large number
of them were built and ran apparently successfully
for a number of years, but on Mr Webb's retirement
they were gradually withdrawn. What economy was
obtained with them remains a secret, but that they
were not good at starting was obvious to all observant

travellers.   The high-pressure cylinders were of
small   diameter   and,   acting   alone,   insufficient to
start a heavy train.   Until, however, they did start
working, the low-pressure cylinder could get no
steam.   What usually happened was that the wheels
driven by the high-pressure cylinder started slipping
badly, giving the large low-pressure cylinder more
than enough steam, so that this started working with
a violent series of jerks which continued during
acceleration, and communicated themselves to the
train, rendering it very uncomfortable for the passengers.

Mr Webb afterwards adopted the four-cylinder
arrangement on his compound engines of the *Black
Prince* type.   These had two outside high-pressure
cylinders, 15 ins. diameter, and two inside low-pressure
cylinders, $16\frac{1}{2}$ ins. in diameter, both pairs having a
common stroke of 24 in.   They were all situated in
line under the smoke-box, and drove on to the first
pair of coupled wheels and its axle.   The wheels were
7 ft. 1 in. diameter.   A number of these engines are
still in service.

Mr Webb's example was followed in 1885 by
Mr James Worsdell on the Gt Eastern Railway, who
used one high-pressure and one low-pressure cylinder
located between the frames. His successor, Mr Holden,
however, converted them all to the ordinary non-
compound type.   On the North Eastern Railway

Mr Worsdell re-introduced the compound system, which his brother, Mr Wilson Worsdell, continued, using the Worsdell-von Borries arrangement of two cylinders. Steam after exhausting from the high-pressure cylinder was passed round the smoke-box to the low-pressure valve chest. A device known as an intercepting valve was introduced on these engines, by which steam could be admitted to the low-pressure cylinder at will by the driver when this was necessary for starting purposes. Both the starting and intercepting valves were operated by steam and controlled by one handle. If the engine did not start when the regulator was opened, which occurred when the engine was 'blinded,' the driver pulled the additional small handle which closed the passage from the receiver* to the low-pressure cylinder, and also admitted a small amount of steam to the low-pressure steam chest, so that the two cylinders together developed additional starting power. After one or two strokes of the engine, the exhaust steam from the high-pressure cylinder automatically forced the two valves back to their normal position, and the engine proceeded, working compound.

With the possible exception of the Midland Rail-

---

* A receptacle used when the cranks are set at 90°. When the h.p. cylinder is exhausting, the port of the h.p. cylinder has not yet been opened for steam. The h.p. exhaust is therefore passed into the receiver from which the h.p. cylinder draws its supply.

way, the subsequent history of compounding in Great
Britain is limited to a series of trial engines on the
Gt Western, Lancashire and Yorkshire, Gt Central
and Gt Northern railways. On the Midland, a
number of three-cylinder engines constructed on the
Smith system are at work. Of the three cylinders,
one is high-pressure and two low-pressure, the high-
pressure cylinder being placed between the frames
and the two low-pressure outside. The high-pressure
cylinder takes steam direct from the boiler, at a
pressure of 220 lbs. per square inch, and this steam
exhausts into the chest common to the low-pressure
cylinders. The steam regulator operates an ingenious
arrangement consisting of a main and jockey valves.
When moved over to start the engine, high-pressure
steam is admitted simultaneously to the main steam
pipe and to the low-pressure auxiliary pipe. When
the main valve is on the point of moving, the area
of the passage by which boiler steam can pass to the
low-pressure cylinder is maximum, and further movement of the handle causes the main valve gradually to
close this opening and also to increase the opening
for the passage of steam from the boiler into the
high-pressure steam pipe. The admission of boiler
steam to the low-pressure cylinder is entirely cut
off by moving the handle about 30° from the shut
position, when the engine of course commences to
work entirely as a compound.

11—2

On the Continent and in America, the compound system has been applied to the locomotive by the following methods:

Two Cylinders, one high-pressure and one low-pressure, driving cranks set at 90°. In normal working one cylinder is supplied with steam at boiler pressure, and an apparatus, sometimes automatically operated, is provided for admitting boiler steam direct to the low-pressure cylinder at starting (Mallet, von Borries, Golsdorf, etc.).

Three Cylinders. This system has found little application abroad.

Four Cylinders. Two high- and two low-pressure cylinders are disposed in tandem in the Woolff type, and operate two cranks set at 90°; two valve gears are employed, the valves in each group being operated by the same link. The addition of a starting apparatus is not indispensable, but generally some provision is made for the direct admission of boiler steam to the low-pressure cylinder. In the Vauclain (America) system, the two high- and two low-pressure cylinders are vertically superposed and drive to a common crosshead. Two valve gears are employed, and one valve distributes steam to each pair of cylinders.

In the *Adriatic* type also, one valve serves each pair of cylinders. The arrangement is very ingenious, the cylinders being grouped as follows:

The two H.P. cylinders are placed on one side of the engine, one inside and one outside, operating two cranks set at 180°; on the other side of the engine are the two L.P. cylinders, one inside and one outside, also driving cranks set at 180° to each other and at 90° to the H.P. pair. The valve is placed above the outside cylinder, and distributes steam to its pair of cylinders by cross passages.

The two high- and two low-pressure cylinders may drive respectively cranks set at 90°, the pairs being at 180° to each other. The H.P. and L.P. cranks are sometimes on the same axle, as in the Maffei and von Borries systems, or drive on two separate coupled axles, as in the famous French type known as the de Glehn and du Bousquet system. This dispenses in principle with the use of a starting apparatus, since the arrangement comprises two high-pressure cylinders with cranks at 90°. In practice, however, it is often applied owing to the small dimensions of the H.P. cylinders which, in certain positions of the cranks, may be unable to start the train. This is the system which has been applied to practically all the French engines constructed since 1896 and with which such excellent results have been obtained, particularly on the Northern of France Railway.

A considerable fuel economy, amounting in some cases to as much as 20 per cent., is definitely admitted to have been found in the working of compound

engines, and as M. Sauvage has pointed out, it is
rather under-estimating the merits of the compounds
to say that by their use the weight of trains is increased by one-third with the same cost of fuel
over what is used with the best simple engines
used before. If not weight, increased speed is
obtained, and in many cases both weight and speed.
His presentation, which represents the French view,
of the case of the compound *versus* simple fairly
considers all the circumstances. He states 'the
initial cost of the compounds is higher, the expenses
for repairs somewhat greater, but the increase of
traffic is such that the economy is obvious. A
complete solution of the problem would require a
proof that the same results might not be obtained
in some other way. Available data are not sufficient
to give such a proof in an incontestable manner;
still, it seems difficult to build an ordinary locomotive
quite equal in every respect to the latest compounds.
It is clear that simple two-cylinder engines might be
made with the same large boiler, and work with the
same high pressure, but it is nearly as clear that,
with the ordinary valve gear of the locomotive,
steam at such a high pressure cannot be utilized
as well as by compounding; there is little doubt
that the simple locomotive would require more
steam for the same work. In addition, there is a
real difficulty in making all the parts of the simple

engine strong enough to stand without undue wear
the greatest stresses resulting from the increased
pressure on large pistons, although this difficulty
may be overcome. An opinion which seems to prevail
is that compound locomotives may be economical
during long runs, but that their advantage is lost
when they stop and start frequently, owing to the
direct admission of steam to the low-pressure
cylinders at starting. This opinion is rather too
dogmatic, and the question requires some consideration. In many cases, with four-cylinder compounds, the tractive power necessary for starting
from rest is obtained without this direct admission,
and steam is admitted in that way only for the first
revolution of wheels. The engine is then worked
compound, but in full gear for all cylinders. Of
course, steam is not so well utilized as with a proper
degree of expansion in each cylinder, but, even in
that case, the compound compares favourably with
a simple locomotive working in full gear.' Under
these circumstances it is difficult to account for the
unpopularity of compounding with British locomotive
engineers. Meagre as they are, the results published
in this country incontestably shew the compound to
be a more efficient and economical machine than the
simple engine. Against this there is to be placed the
statement made in some quarters, that the additional
cost of maintenance, due to increased complication,

more than neutralizes the advantages gained in fuel saving. It is difficult to see, however, that the adoption of three or four cylinders working non-compound does not introduce the same increase in cost of maintenance ; which may also be said of the addition of superheating apparatus to engines of the ordinary two-cylinder type involving the installation of mechanical lubricators, piston valves and numerous accessories. M. Demoulin, of the Western Railway of France, has even stated that the capital involved is more than that required for compounding, for the same number of cylinders ; and it will probably be admitted that superheating when applied under the most favourable conditions, although yielding an economy analogous to that which results from compounding, is nevertheless inferior thereto.

This does not mean that continental engineers are neglecting superheating ; on the other hand they are using it in combination with compounding, which seems to be highly advantageous if it can be obtained without adding greatly to the complexity of the whole machine. It is difficult to reconcile conflicting opinions, hence the writer has limited himself to a simple statement of the question.

The Future. Turbine locomotives have been experimented with in Germany and by Mr Reid, of the North British Locomotive Co., but have not met with much success. The efficiency of the turbine depends essentially upon adequate condensing arrangements, for which air cooling or the limited quantity of cooling water capable of being carried on an engine is quite inadequate. The future may see an increased application of the water pick-up system, which would considerably advance matters in this direction, as not only would the necessary partial vacuum then be maintained, but the condensed steam could be pumped back into the boiler at a high temperature. The provision of such facilities would equally favour compounding, and enable results to be obtained therefrom comparable with those realised in marine and stationary practice. On the other hand no exhaust steam would be available for the purpose of creating a draught.

BIBLIOGRAPHY

LAKE. The Locomotive Simply Explained. Percival Marshall & Co. A useful introduction to the subject.

HUGHES. The Construction of the Modern Locomotive. E. and F. N. Spon. A practical work dealing with methods of manufacture and erection.

PETTIGREW. A Manual of Locomotive Engineering. Chas. Griffin & Co. The standard English work.

ANON. The Locomotive of To-day. The Locomotive Publishing Co. An excellent work dealing very fully with constructional details.

DEMOULIN. 1. Traité de la Machine Locomotive. 2. La Locomotive Actuelle. Béranger, Paris. A standard work.

NADAL. Locomotives à Vapeur. Octave Doin, Paris. Very valuable for its clear mathematical treatment of many problems untouched in other treatises.

GARBE. Die Dampflokomotiven der Gegenwart. Julius Springer, Berlin. The standard German treatise.

PENDRED. The Railway Locomotive. Constable & Co. A very attractive work, containing much useful information not found elsewhere. Treatment not too technical.

DALBY. The Balancing of Engines. Edwin Arnold & Co.

VON BORRIES. Die Lokomotiven der Gegenwart. Kreidel, Wiesbaden.

PAPERS AND ARTICLES.

ASPINALL. Train Resistance. Proceedings Institution Civil Engineers. 1901–2.

MARSHALL. The Evolution of the Locomotive Engine. Proceedings Institution Civil Engineers. 1897–8.

SAUVAGE. Recent Locomotive Practice in France. Proceedings Institution Mechanical Engineers. 1900.

CHURCHWARD. Large Locomotive Boilers. Proceedings Institution Mechanical Engineers. 1906.

HUGHES. Locomotives designed and built at Horwich with some Results. Proceedings Institution Mechanical Engineers. 1909.

—— Compounding and Superheating in Horwich Locomotives. Proceedings Institution Mechanical Engineers. 1910.

SAMS. Modern Locomotive Construction. The Engineering Review. 1908.

SUMNER. The Power of a Locomotive Boiler. The Engineering Review. 1910.

—— Coal Consumption on Locomotives. The Engineering Review. 1909.

STROUDLEY. The Construction of Locomotive Engines. Proceedings Institution Civil Engineers. 1885.

INDEX

CPSIA information can be obtained at www.ICGtesting.com
Printed in the USA
LVOW06s0311131215

466420LV00001B/84/P